# Design Examples for High Strength Steel Reinforced Concrete Columns

## A Eurocode 4 Approach

T0177128

# Design Examples for High Strength Steel Reinforced Concrete Columns

## A Eurocode 4 Approach

Sing-Ping Chiew
Yan-Qing Cai

CRC Press
Taylor & Francis Group
Boca Raton London New York

CRC Press is an imprint of the
Taylor & Francis Group, an **Informa** business

CRC Press
Taylor & Francis Group
6000 Broken Sound Parkway NW, Suite 300
Boca Raton, FL 33487-2742

First issued in paperback 2021

Printed on acid-free paper

ISBN-13: 978-1-138-60269-4 (hbk)
ISBN-13: 978-1-03-209558-5 (pbk)

### Library of Congress Cataloging-in-Publication Data

Names: Chiew, Sing-Ping, author. | Cai, Y. Q. (Yan Qing), author.
Title: Design of high strength steel reinforced concrete columns : a Eurocode 4 approach / S.P. Chiew and Y.Q. Cai.
Description: Boca Raton : CRC Press, [2018] | Includes bibliographical references and index.
Identifiers: LCCN 2017057555 (print) | LCCN 2018000768 (ebook) | ISBN 9781351203944 (Adobe PDF) | ISBN 9781351203937 (ePub) | ISBN 9781351203920 (Mobipocket) | ISBN 9780815384601 (hardback : acid-free paper) | ISBN 9781351203951 (ebook)
Subjects: LCSH: Composite construction--Specifications--Europe. | Building, Iron and steel--Specifications--Europe. | Reinforced concrete construction--Specifications--Europe.
Classification: LCC TA664 (ebook) | LCC TA664 .C48 2018 (print) | DDC 624.1/83425--dc23
LC record available at https://lccn.loc.gov/2017057555

**Visit the Taylor & Francis Web site at**
**http://www.taylorandfrancis.com**

**and the CRC Press Web site at**
**http://www.crcpress.com**

# Contents

# List of symbols

$A_a$        Area of the structural steel
$A_c$        Area of concrete
$A_{ch}$      Area of highly confined concrete
$A_{cp}$      Area of partially confined concrete
$A_{cu}$      Area of unconfined concrete
$A_s$        Area of reinforcement
$E_a$        Modulus of elasticity of structural steel
$E_{c,eff}$    Effective modulus of elasticity of concrete
$E_{cm}$      Secant modulus of elasticity of concrete
$E_c(t)$      Tangent modulus of elasticity of concrete at time $t$
$E_s$        Modulus of elasticity of reinforcement
$(EI)_{eff}$   Effective flexural stiffness
$G_a$        Shear modulus of structural steel
$I$          Second moment of area of the composite section
$I_a$        Second moment of area of the structural steel
$I_c$        Second moment of area of the concrete
$I_s$        Second moment of area of the reinforcement
$K_e$        Correction factor
$L$          Length
$M_{Ed}$      Design bending moment
$M_{pl,a,Rd}$  The plastic resistance moment of the structural steel
$M_{pl,Rd}$    The plastic resistance moment of the composite section
$N_{cr}$      Elastic critical force in composite columns
$N_{Ed}$      The compressive normal force
$N_{pl,Rd}$    The plastic resistance of the composite section
$N_{pl,Rk}$    Characteristic value of the plastic resistance of the
             composite section
$N_{pm,Rd}$    The resistance of the concrete to compressive normal force
$P_{Rd}$      The resistance of per shear stud

| | |
|---|---|
| $V_{\mathrm{Ed}}$ | The shear force |
| $V_{\mathrm{pl,a,Rd}}$ | The shear resistance of the steel section |
| $W_{\mathrm{pa}}$ | The plastic section modulus of the structural steel |
| $W_{\mathrm{pc}}$ | The plastic section modulus of the concrete |
| $W_{\mathrm{ps}}$ | The plastic section modulus of the reinforcing steel |
| $b_{\mathrm{c}}$ | Width of the composite section |
| $b_{\mathrm{f}}$ | Width of the steel flange |
| $c_{\mathrm{y}}, c_{\mathrm{z}}$ | Thickness of concrete cover |
| $d$ | Diameter of shank of the headed stud |
| $e$ | Eccentricity of loading |
| $f_{\mathrm{ck}}$ | The cylinder compressive strength of concrete |
| $f_{\mathrm{cd}}$ | The design strength of concrete |
| $f_{\mathrm{c,p}}$ | The compressive strength of partially confined concrete |
| $f_{\mathrm{c,h}}$ | The compressive strength of highly confined concrete |
| $f_{\mathrm{s}}$ | The yield strength of reinforcement |
| $f_{\mathrm{u}}$ | Tensile strength |
| $f_{\mathrm{y}}$ | The yield strength of structural steel |
| $f_{\mathrm{yd}}$ | The design strength of structural steel |
| $f_{\mathrm{yh}}$ | The yield strength of transverse reinforcement |
| $h_{\mathrm{a}}$ | Depth of steel section |
| $h_{\mathrm{c}}$ | Depth of composite section |
| $h_{\mathrm{n}}$ | Distance from centroidal axis to neutral axis |
| $h_{\mathrm{sc}}$ | Overall nominal height of the headed stud |
| $s$ | Spacing center-to-center of links |
| $t_{\mathrm{f}}$ | Thickness of steel flange |
| $t_{\mathrm{w}}$ | Thickness of the steel web |
| $\Delta\sigma$ | Stress range |
| $\Psi$ | Coefficient |
| $\alpha$ | Coefficient; factor |
| $\beta$ | Factor; coefficient |
| $\gamma$ | Partial factor |
| $\delta$ | Steel contribution ratio |
| $\eta$ | Coefficient |
| $\varepsilon_{\mathrm{c,u}}$ | Strain of unconfined concrete |
| $\varepsilon_{\mathrm{c,p}}$ | Strain of partially confined concrete |
| $\varepsilon_{\mathrm{c,h}}$ | Strain of highly confined concrete |
| $\mu$ | Factors related to bending moments |
| $\lambda$ | Relative slenderness |
| $\rho_{\mathrm{s}}$ | Reinforcement ratio |
| $\chi$ | Reduction factor of buckling |
| $\varphi$ | Creep coefficient |

# Preface

This book is the companion volume to *Design of High Strength Steel Reinforced Concrete Columns—A Eurocode 4 Approach*.

Guidance is much needed on the design of high strength steel reinforced concrete (SRC) columns beyond the remit of Eurocode 4 for composite steel concrete structures. Given the much narrower range of permitted concrete and steel material strengths in comparison to Eurocode 2 for concrete structures and Eurocode 3 for steel structures, and the better ductility and buckling resistance of SRC columns compared to steel or reinforced concrete, there is a clear need for design beyond the current guidelines. The design principles to do so are set out in the companion volume to this book, *Design of High Strength Steel Reinforced Concrete Columns—A Eurocode 4 Approach*. This book provides a number of design examples for high strength SRC columns using these principles which are based on the Eurocode 4 approach. Special considerations are given to resistance calculations that maximize the full strength of the materials, with concrete cylinder strength up to $90\,N/mm^2$, yield strength of structural steel up to $690\,N/mm^2$ and yield strength of reinforcing steel up to $600\,N/mm^2$ respectively. These design examples will allow the readers to practice and understand the Eurocode 4 methodology easily.

Structural engineers and designers who are familiar with basic Eurocode 4 design should find these design examples particularly helpful, whilst civil engineering students who are studying composite steel concrete design and construction should gain further understanding from working through the design examples which are set out clearly in a step-by-step fashion.

# Authors

**Sing-Ping Chiew** is a professor and the Civil Engineering Programme director at the Singapore Institute of Technology, Singapore, and coauthor of *Structural Steelwork Design to Limit State Theory, 4th Edition.*

**Yan-Qing Cai** is a project officer in the School of Civil and Environmental Engineering at Nanyang Technological University, Singapore.

# Design examples

## STEEL-REINFORCED CONCRETE COLUMN SUBJECTED TO AXIAL COMPRESSION

Determine the axial buckling resistance of SRC columns (concrete-encased I-section) subject to pure compression with an effective length of 4 m, as shown in Figure 1.

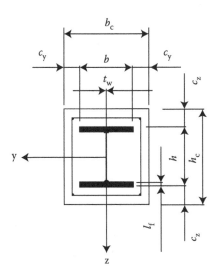

*Figure 1* Cross-section of SRC column.

## Steel-reinforced concrete column with normal-strength material

### Design data

| | |
|---|---|
| Design axial force | $N_{Ed} = 8000$ kN |
| Permanent load | $N_{G,Ed} = 4000$ kN |
| Column length | $L = 4.0$ m |
| Effective length | $L_{eff} = 4.0$ m |
| Structural steel | Grade S355, $f_y = 355$ N/mm² |
| Concrete | C30/37, $f_{ck} = 30$ N/mm² |
| Reinforcement | $f_{sk} = 500$ N/mm² |

**Properties of cross-section**

| | |
|---|---|
| Concrete depth | $h_c = 500$ mm |
| Concrete width | $b_c = 500$ mm |
| Concrete cover | $c = 30$ mm |
| Cover | $c_y = 95.4$ mm |
| Cover | $c_z = 89.8$ mm |

**Section properties of steel section**
305 × 305 UC 137

| | |
|---|---|
| Depth | $h = 320.5$ mm |
| Width | $b = 309.2$ mm |
| Flange thickness | $t_f = 21.7$ mm |
| Web thickness | $t_w = 13.8$ mm |
| Fillet | $r = 15.2$ mm |
| Section area | $A_a = 174.4$ cm² |
| Second moment of area/y | $I_{ay} = 32810$ cm⁴ |
| Second moment of area/z | $I_{az} = 10700$ cm⁴ |
| Plastic section modulus/y | $W_{pl,a,y} = 2297$ cm³ |
| Plastic section modulus/z | $W_{pl,a,z} = 1053$ cm³ |

**Reinforcement**

| | |
|---|---|
| Longitudinal reinforcement | number $n = 8$, diameter $d_{l,s} = 20$ mm |
| Transverse reinforcement | diameter $d_{t,s} = 10$ mm, spacing $s = 200$ mm |

### Design strengths and modulus

According to Tables 2.12 and 2.13 in the companion book, for the strain compatibility between steel and concrete, the two materials (concrete class C30/37 and steel grade S355) are strain compatible. Therefore, the steel section can reach its full strength when the composite concrete section reaches its ultimate strength, without considering the confinement effect from the lateral hoops and steel section.

The design strengths of the steel, concrete, and reinforcement are:

$$f_{yd} = \frac{f_y}{\gamma_M} = \frac{355}{1.0} = 355 \, N/mm^2$$

$$f_{cd} = \frac{f_{ck}}{\gamma_C} = \frac{30}{1.5} = 20 \, N/mm^2$$

$$f_{sd} = \frac{f_{sk}}{\gamma_S} = \frac{500}{1.15} = 435 \, N/mm^2$$

$$E_{cm} = 33 \, Gpa$$

$$E_a = 210 \, Gpa$$

### Cross-sectional areas

The cross-sectional areas of the steel, reinforcement, and concrete are:

$$A_a = 17,440 \, mm^2$$

$$A_s = \frac{8 \times \pi \times 20^2}{4} = 2512 \, mm^2$$

$$A_c = b_c h_c - A_a - A_s = 500 \times 500 - 17,440 - 2512 = 230,048 \, mm^2$$

### Second moments of area

$$I_{ay} = 328.1 \times 10^6 \, mm^4$$

$$I_{az} = 107 \times 10^6 \, mm^4$$

$$I_{sy} = \sum_{i=1}^{n} A_{s,i} e_i^2 = 6 \times \frac{\pi \times 20^4}{4} \times 200^2 = 73.56 \times 10^6 \, mm^4$$

$$I_{sz} = \sum_{i=1}^{n} A_{s,i}e_i^2 = 6 \times \frac{\pi \times 20^4}{4} \times 200^2 = 73.56 \times 10^6 \, \text{mm}^4$$

$$I_{cy} = \frac{b_c h_c^3}{12} - I_{ay} - I_{sy}$$
$$= \frac{500 \times 500^3}{12} - 328.1 \times 10^6 - 73.56 \times 10^6 = 4805 \times 10^6 \, \text{mm}^4$$

$$I_{cz} = \frac{h_c b_c^3}{12} - I_{az} - I_{sz}$$
$$= \frac{500 \times 500^3}{12} - 107 \times 10^6 - 73.56 \times 10^6 = 5026 \times 10^6 \, \text{mm}^4$$

### Check the reinforcement ratio

$$\rho_s = \frac{A_s}{A_c} = \frac{2512}{230,048} = 1.1\%$$

The reinforcement ratio is within the range 0.3%–6%.

### Check the local buckling

The concrete cover to the flange of the steel section: $c = 89.8$ mm > maximum (40 mm; $b_f/6$).

Thus, the effect of local buckling is neglected for the SRC column.

### Check the steel contribution factor

The design plastic resistance of the composite cross-section in compression is:

$$N_{pl,Rd} = A_a f_{yd} + 0.85 A_c f_{cd} + A_s f_{sd}$$
$$= (17,440 \times 355 + 0.85 \times 230,048 \times 20 + 2512 \times 435) \times 10^{-3}$$
$$= 11,195 \, \text{kN}$$

$$\delta = \frac{A_a f_{yd}}{N_{pl,Rd}} = \frac{17,440 \times 355 \times 10^{-3}}{11,195} = 0.55$$

which is within the permitted range, $0.2 \leq \delta \leq 0.9$.

## Long-term effects

The age of concrete at loading $t_0$ is assumed to be 14 days. The age of concrete at the moment considered $t$ is taken as infinity. The relative humidity $RH$ is taken as 50%.

The notional size of the cross-section is:

$$h_0 = \frac{2A_c}{u} = \frac{2 \times 230,048}{4 \times 500} = 230 \, \text{mm}$$

Coefficient:

$$\alpha_1 = \left(\frac{35}{f_{cm}}\right)^{0.7} = \left(\frac{35}{30+8}\right)^{0.7} = 0.94$$

$$\alpha_2 = \left(\frac{35}{f_{cm}}\right)^{0.2} = \left(\frac{35}{38}\right)^{0.2} = 0.98$$

$$\alpha_3 = \left(\frac{35}{f_{cm}}\right)^{0.5} = \left(\frac{35}{38}\right)^{0.5} = 0.96$$

Factor:

$$\varphi_{RH} = \left[1 + \frac{1 - RH/100}{0.1\sqrt[3]{h_0}} \alpha_1\right] \alpha_2 = \left[1 + \frac{1 - 50/100}{0.1\sqrt[3]{230}} \times 0.94\right] \times 0.98 = 1.73$$

$$\beta(f_{cm}) = \frac{16.8}{\sqrt{f_{cm}}} = \frac{16.8}{\sqrt{38}} = 2.73$$

$$\beta(t_0) = \frac{1}{(0.1 + t_0^{0.20})} = \frac{1}{(0.1 + 14^{0.20})} = 0.56$$

$$\varphi_0 = \varphi_{RH}\beta(f_{cm})\beta(t_0) = 1.73 \times 2.73 \times 0.56 = 2.64$$

Factor:

$$\beta_H = 1.5[1 + (0.012 RH)^{18}]h_0 + 250\alpha_3$$
$$= 1.5[1 + (0.012 \times 50)^{18}] \times 230 + 250 \times 0.96 = 585$$

$$\beta(t,t_0) = \left[\frac{t - t_0}{(\beta_H + t - t_0)}\right]^{0.3} = \left[\frac{\infty - 14}{(585 + \infty - 14)}\right]^{0.3} = 1.0$$

The creep coefficient is:

$$\varphi_t = \varphi_0\beta_c(t,t_0) = 2.64 \times 1.0 = 2.64$$

### Elastic modulus of concrete considering long-term effects

Long-term effects due to creep and shrinkage should be considered in determining the effective elastic flexural stiffness. The modulus of elasticity of concrete $E_{cm}$ is reduced to the value $E_{c,eff}$:

$$E_{c,eff} = \frac{E_{cm}}{1 + (N_{G,Ed}/N_{Ed})\varphi_t} = \frac{33}{1 + (4000/8000) \times 2.64} = 14.22 \, \text{kN/mm}^2$$

### Effective flexural stiffness of cross-section

The effective elastic flexural stiffness taking account of the long-term effects is:

$$(EI)_{eff,y} = E_a I_{ay} + 0.6 E_{c,eff} I_{cy} + E_s I_{sy}$$
$$= 210 \times 328.1 \times 10^6 + 0.6 \times 14.22 \times 4805 \times 10^6 + 210 \times 73.56 \times 10^6$$
$$= 1.25 \times 10^{11} \, \text{kN mm}^2$$

$$(EI)_{eff,z} = E_a I_{az} + 0.6 E_{c,eff} I_{cz} + E_s I_{sz}$$
$$= 210 \times 107 \times 10^6 + 0.6 \times 14.22 \times 5026 \times 10^6 + 210 \times 73.56 \times 10^6$$
$$= 8.04 \times 10^{10} \, \text{kN mm}^2$$

## Elastic critical normal force

$$N_{cry} = \frac{\pi^2 (EI)_{eff,y}}{L_y^2} = \frac{\pi^2 \times 1.25 \times 10^{11}}{4^2 \times 10^6} = 77,000\,\text{kN}$$

$$N_{crz} = \frac{\pi^2 (EI)_{eff,z}}{L_z^2} = \frac{\pi^2 \times 8.04 \times 10^{10}}{4^2 \times 10^6} = 49,600\,\text{kN}$$

The characteristic value of the plastic resistance to the axial load is:

$$\begin{aligned} N_{pl,Rk} &= A_a f_y + 0.85 A_c f_{ck} + A_s f_{sk} \\ &= (17,440 \times 355 + 0.85 \times 230,048 \times 30 + 2512 \times 500) \times 10^3 \\ &= 13,313\,\text{kN} \end{aligned}$$

## Relative slenderness ratio

$$\overline{\lambda_y} = \sqrt{\frac{N_{pl,Rk}}{N_{cry}}} = \sqrt{\frac{13,313}{77,000}} = 0.42$$

$$\overline{\lambda_z} = \sqrt{\frac{N_{pl,Rk}}{N_{crz}}} = \sqrt{\frac{13,313}{49,600}} = 0.52$$

The nondimensional slenderness does not exceed 2.0, so the simplified design method is applicable.

## Buckling reduction factor

Buckling curve $b$ is applicable to axis $y$-$y$, and buckling curve $c$ is applicable to axis $z$-$z$ in accordance with EN 1994-1-1. The imperfection factor is taken as 0.34 for curve $b$ and 0.49 for curve $c$. According to EN 1993-1-1, the factor is:

$$\begin{aligned} \Phi_y &= 0.5 \left( 1 + \alpha \left( \overline{\lambda_y} - 0.2 \right) + \overline{\lambda_y^2} \right) \\ &= 0.5 \times [1 + 0.34 \times (0.42 - 0.2) + 0.42^2] = 0.62 \end{aligned}$$

$$\Phi_z = 0.5\left(1 + \alpha\left(\overline{\lambda_z} - 0.2\right) + \overline{\lambda_z^2}\right)$$
$$= 0.5 \times [1 + 0.49 \times (0.52 - 0.2) + 0.52^2] = 0.71$$

The reduction factor for column buckling is:

$$\chi_y = \frac{1}{\Phi_y + \sqrt{\Phi_y^2 - \overline{\lambda_y^2}}} = \frac{1}{0.62 + \sqrt{0.62^2 - 0.42^2}} = 0.92$$

$$\chi_z = \frac{1}{\Phi_z + \sqrt{\Phi_z^2 - \overline{\lambda_z^2}}} = \frac{1}{0.71 + \sqrt{0.71^2 - 0.52^2}} = 0.83$$

### Buckling resistance

The minor axis is the more critical, so

$$N_{b,Rd} = \min(\chi_y; \chi_z)N_{pl,Rd}$$
$$= 0.83 \times 11{,}195 = 9292\,\text{kN} > N_{Ed} = 8000\,\text{kN}$$

The buckling resistance of the SRC column is adequate.

## Steel-reinforced concrete column with high-strength concrete

Concrete class C90/105 is used in this design example. Other design data are same as in Section "Steel-reinforced concrete column with normal-strength material," such as loading; column length; steel strength; and the dimensions of the SRC column cross-section, steel section, and reinforcement.

### Design strengths and modulus

According to Tables 2.12 and 2.13 in the companion book, for the strain compatibility between steel and concrete, the two materials (concrete class C90/105 and steel grade S355) are strain compatible, so the steel can reach its full strength when the composite concrete section reaches its ultimate strength without considering the confinement effect from the lateral hoops and steel section.

For high-strength concrete with $f_{ck} > 50\,\text{N/mm}^2$, the effective compressive strength of concrete should be used in accordance with EC2. The effective strength is:

$$f_{ck} = 90\eta = 90 \times (1.0 - (90 - 50)/200) = 72\,\text{N/mm}^2$$

Accordingly, the secant modulus for high-strength concrete C90/105 is

$$E_{cm} = 22[(\eta f_{ck} + 8)/10]^{0.3} = 22[(72 + 8)/10]^{0.3} = 41.1\,\text{GPa}$$

Then, the design strengths of the steel, concrete, and reinforcement are:

$$f_{yd} = \frac{f_y}{\gamma_M} = \frac{355}{1.0} = 355\,\text{N/mm}^2$$

$$f_{cd} = \frac{f_{ck}}{\gamma_C} = \frac{72}{1.5} = 48\,\text{N/mm}^2$$

$$f_{sd} = \frac{f_{sk}}{\gamma_S} = \frac{500}{1.15} = 435\,\text{N/mm}^2$$

### Cross-sectional areas and second moments of area

The cross-sectional area and second moment area of the steel, reinforcement, and concrete are the same as the design example in Section "Steel-reinforced concrete column with normal-strength material."

$A_a = 17{,}440\,\text{mm}^2$, $A_s = 2512\,\text{mm}^2$, $A_c = 230{,}048\,\text{mm}^2$
$I_{ay} = 328.1 \times 10^6\,\text{mm}^4$, $I_{az} = 107 \times 10^6\,\text{mm}^4$
$I_{sy} = 73.56 \times 10^6\,\text{mm}^4$, $I_{sz} = 73.56 \times 10^6\,\text{mm}^4$
$I_{cy} = 4805 \times 10^6\,\text{mm}^4$, $I_{cz} = 5026 \times 10^6\,\text{mm}^4$

### Check the steel contribution factor

The design plastic resistance of the composite cross-section in compression is:

$$
\begin{aligned}
N_{pl,Rd} &= A_a f_{yd} + 0.85 A_c f_{cd} + A_s f_{sd} \\
&= (17{,}440 \times 355 + 0.85 \times 230{,}048 \times 48 + 2512 \times 435) \times 10^{-3} \\
&= 16{,}669\,\text{kN}
\end{aligned}
$$

$$\delta = \frac{A_a f_{yd}}{N_{pl,Rd}} = \frac{17,440 \times 355 \times 10^{-3}}{16,669} = 0.37$$

which is within the permitted range, $0.2 \leq \delta \leq 0.9$.

### Long-term effects

The age of concrete at loading $t_0$ is assumed to be 14 days. The age of concrete at the moment considered $t$ is taken as infinity. The relative humidity $RH$ is taken as 50%.

The notional size of the cross-section is:

$$h_0 = 2A_c/u = 230 \text{ mm}$$

Coefficient:

$$\alpha_1 = \left(\frac{35}{f_{cm}}\right)^{0.7} = \left(\frac{35}{72+8}\right)^{0.7} = 0.56$$

$$\alpha_2 = \left(\frac{35}{f_{cm}}\right)^{0.2} = \left(\frac{35}{80}\right)^{0.2} = 0.85$$

$$\alpha_3 = \left(\frac{35}{f_{cm}}\right)^{0.5} = \left(\frac{35}{80}\right)^{0.5} = 0.66$$

Factor:

$$\varphi_{RH} = \left[1 + \frac{1 - RH/100}{0.1\sqrt[3]{h_0}}\alpha_1\right]\alpha_2 = \left[1 + \frac{1 - 50/100}{0.1\sqrt[3]{230}} \times 0.56\right] \times 0.85 = 1.24$$

$$\beta(f_{cm}) = \frac{16.8}{\sqrt{f_{cm}}} = \frac{16.8}{\sqrt{80}} = 1.88$$

$$\beta(t_0) = \frac{1}{(0.1 + t_0^{0.20})} = \frac{1}{(0.1 + 14^{0.20})} = 0.56$$

$$\varphi_0 = \varphi_{RH}\beta(f_{cm})\beta(t_0) = 1.24 \times 1.88 \times 0.56 = 1.3$$

Factor:

$$\beta_H = 1.5\,[1+(0.012\mathrm{RH})^{18}]\,h_0 + 250\alpha_3$$
$$= 1.5\,[1+(0.012\times50)^{18}]\times230+250\times0.66 = 510$$

$$\beta(t,t_0) = \left[\frac{t-t_0}{(\beta_H + t - t_0)}\right]^{0.3} = \left[\frac{\infty-14}{(510+\infty-14)}\right]^{0.3} = 1.0$$

The creep coefficient is:

$$\varphi_t = \varphi_0\beta_c(t,t_0) = 1.30\times1.0 = 1.30$$

### Elastic modulus of concrete considering long-term effects

Long-term effects due to creep and shrinkage should be considered in determining the effective elastic flexural stiffness. The modulus of elasticity of concrete $E_{cm}$ is reduced to the value $E_{c,eff}$:

$$E_{c,eff} = \frac{E_{cm}}{1+(N_{G,Ed}/N_{Ed})\varphi_t} = \frac{41.1}{1+(4000/8000)\times1.3} = 24.9\,\mathrm{kN/mm^2}$$

### Effective flexural stiffness of cross-section

$$(EI)_{eff,y} = E_aI_{ay} + 0.6E_{c,eff}I_{cy} + E_sI_{sy}$$
$$= 210\times328.1\times10^6 + 0.6\times24.9\times4805\times10^6 + 210\times73.56\times10^6$$
$$= 1.56\times10^{11}\,\mathrm{kNmm^2}$$

$$(EI)_{eff,z} = E_aI_{az} + 0.6E_{c,eff}I_{cz} + E_sI_{sz}$$
$$= 210\times107\times10^6 + 0.6\times24.9\times5026\times10^6 + 210\times73.56\times10^6$$
$$= 1.13\times10^{11}\,\mathrm{kN\,mm^2}$$

### Elastic critical normal force

$$N_{cry} = \frac{\pi^2(EI)_{eff,y}}{L_y^2} = \frac{\pi^2\times1.56\times10^{11}}{4^2\times10^6} = 96,100\,\mathrm{kN}$$

$$N_{crz} = \frac{\pi^2 (EI)_{eff,z}}{L_z^2} = \frac{\pi^2 \times 1.13 \times 10^{11}}{4^2 \times 10^6} = 69,500 \, \text{kN}$$

The characteristic value of the plastic resistance to the axial load is:

$$N_{pl,Rk} = A_a f_y + 0.85 A_c f_{ck} + A_s f_{sk}$$
$$= (17,440 \times 355 + 0.85 \times 230,048 \times 72 + 2512 \times 500) \times 10^3$$
$$= 21,526 \, \text{kN}$$

### *Relative slenderness ratio*

$$\overline{\lambda_y} = \sqrt{\frac{N_{pl,Rk}}{N_{cry}}} = \sqrt{\frac{21,526}{96,100}} = 0.47$$

$$\overline{\lambda_z} = \sqrt{\frac{N_{pl,Rk}}{N_{crz}}} = \sqrt{\frac{21,526}{69,500}} = 0.56$$

### *Buckling reduction factor*

Buckling curve *b* is applicable to axis *y-y*, and buckling curve *c* is applicable to axis *z-z* in accordance with EN 1994-1-1. The imperfection factor is taken as 0.34 for curve *b* and 0.49 for curve *c*.

$$\Phi_y = 0.5 \left( 1 + \alpha \left( \overline{\lambda_y} - 0.2 \right) + \overline{\lambda_y^2} \right)$$
$$= 0.5 \times [1 + 0.34 \times (0.47 - 0.2) + 0.47^2] = 0.66$$

$$\Phi_z = 0.5 \left( 1 + \alpha \left( \overline{\lambda_z} - 0.2 \right) + \overline{\lambda_z^2} \right)$$
$$= 0.5 \times [1 + 0.49 \times (0.56 - 0.2) + 0.56^2] = 0.74$$

The reduction factor for column buckling is:

$$\chi_y = \frac{1}{\Phi_y + \sqrt{\Phi_y^2 - \overline{\lambda_y^2}}} = \frac{1}{0.66 + \sqrt{0.66^2 - 0.47^2}} = 0.90$$

$$\chi_z = \frac{1}{\Phi_z + \sqrt{\Phi_z^2 - \lambda_z^2}} = \frac{1}{0.74 + \sqrt{0.74^2 - 0.56^2}} = 0.81$$

### Buckling resistance

The minor axis is the more critical, so

$$N_{b,Rd} = \min(\chi_y; \chi_z)N_{pl,Rd}$$
$$= 0.81 \times 16,669 = 13,502 \, \text{kN} > N_{Ed} = 8000 \, \text{kN}$$

The buckling resistance of the SRC column is adequate.

Compared to the SRC column with concrete class C30/37, the buckling resistance ratio is:

$$\frac{N_{b,Rd,C90/105}}{N_{b,Rd,C30/37}} = \frac{13,502}{9292} = 1.45$$

The buckling resistance of an SRC column with high-strength concrete C90/105 is increased by 45% compared to the resistance of a column with C30/37 concrete.

## Steel-reinforced concrete column with high-strength steel

Steel grade S550 is used in this design example. Other design data are same as in Section "Steel-reinforced concrete column with normal-strength material," such as loading; column length; concrete strength; and dimensions of the SRC column cross-section, steel section, and reinforcement.

### Design strengths and modulus

According to Tables 2.12 and 2.13 in the companion book, for the strain compatibility between steel and concrete, the two materials (concrete class C30/37 and steel grade S550) are not strain compatible, so the high-strength concrete reaches its peak strain much earlier than the yield strain of steel. This implies that the concrete will fail earlier than the steel, resulting in a partial utilization of the steel strength. Using the strain-compatibility method, the strength of steel is limited to the stress corresponding to the crushing strain of concrete. The confinement effect from the lateral hoops and steel section is considered as follows.

Longitudinal reinforcement ratio:

$$\rho_s = \frac{A_s}{A_c} = 1.1\%$$

Factor:

$$k_e = \frac{\left(1 - \sum_{i=1}^{n}((b_i)^2/6b_c h_c)\right)(1-(s/2b_c))(1-(s/2h_c))}{1-\rho_s} = 0.514$$

The effective volume ratio of the hoops is:

$$\rho_{se} = k_e \rho_{s,h} = 0.514 \times 0.2\% = 0.1\%$$

The real stress of the hoops is calculated by the modified confinement model:

$$\kappa = \frac{f_{c,u}}{\rho_{se} E_s \varepsilon_c} = \frac{30}{0.001 \times 210 \times 0.0022} = 65$$

$$f_{r,h} = \max\left(\frac{0.25 f_{c,u}}{\rho_{se}(\kappa-10)}; 0.43\varepsilon_c E_s\right)$$

$$= \max\left(\frac{0.25 \times 30}{0.001(65-10)}; 0.43 \times 0.0022 \times 210{,}000\right) = 199\,\text{N/mm}^2$$

The effective lateral confining pressure for PCC from the hoops is:

$$f_{l,p} = \rho_{se} f_{r,h} = 0.001 \times 199 = 0.199\,\text{N/mm}^2$$

The strain of PCC is:

$$\varepsilon_{c,p} = \left[1 + 35\left(\frac{f_{l,p}}{f_{c,u}}\right)^{1.2}\right]\varepsilon_c = \left[1 + 35\left(\frac{0.199}{30}\right)^{1.2}\right] \times 0.0022 = 0.0024$$

Factor:

$$k_e' = \frac{A_{c,f} - A_{c,r}}{A_{c,f}} = \frac{40{,}927 - 12{,}797}{40{,}927} = 0.69$$

Factor:

$$k_a = \frac{t_f^2}{3l^2} = 0.0072$$

The effective lateral confining pressure from the steel section is:

$$f_{l,s} = k_e' k_a f_{r,y} = 2.5\,\text{N/mm}^2$$

The effective lateral confining stress for HCC is:

$$f_{l,h} = f_{l,p} + f_{l,s} = 0.199 + 2.5 = 2.699\,\text{N/mm}^2$$

The strain of HCC is:

$$\varepsilon_{c,p} = \left(1 + 35\left(\frac{f_{l,h}}{f_{c,u}}\right)^{1.2}\right)\varepsilon_c = \left(1 + 35\left(\frac{2.699}{30}\right)^{1.2}\right)\times 0.0022 = 0.006$$

To ensure the yield strain of steel is less than the compressive strain of concrete, the maximum steel strength can be determined accordingly. The real stress of the steel flange in partially confined concrete is:

$$f_{r,f} = \varepsilon_{c,p}E_a = 0.0024 \times 210{,}000 = 504\,\text{N/mm}^2$$

The real stress of the steel web in highly confined concrete is:

$$f_{r,w} = \min(\varepsilon_{c,h}E_a; f_y) = \min(0.006 \times 210{,}000; 550) = 550\,\text{N/mm}^2$$

The steel strength in partially confined concrete is lower than the yield strength of steel, 550 N/mm². The confinement pressure is insufficient to ensure the utilization of steel's full strength. A higher confinement level is needed. Thus, the conservative value of the steel flange is taken as the steel strength in the following design.

Then, the design strength of steel is:

$$f_{yd} = \frac{f_y}{\gamma_M} = \frac{504}{1.0} = 504\,\text{N/mm}^2$$

### Cross-sectional areas and second moments of area

The cross-sectional area and second moment area of the steel, reinforcement, and concrete are the same as the design example in Section "Steel-reinforced concrete column with normal-strength material."

$A_a = 17{,}440$ mm$^2$, $A_s = 2512$ mm$^2$, $A_c = 230{,}048$ mm$^2$
$I_{ay} = 328.1 \times 10^6$ mm$^4$, $I_{az} = 107 \times 10^6$ mm$^4$
$I_{sy} = 73.56 \times 10^6$ mm$^4$, $I_{sz} = 73.56 \times 10^6$ mm$^4$
$I_{cy} = 4805 \times 10^6$ mm$^4$, $I_{cz} = 5026 \times 10^6$ mm$^4$

### Check the steel contribution factor

The design plastic resistance of the composite cross-section in compression is:

$$N_{pl,Rd} = A_a f_{yd} + 0.85 A_c f_{cd} + A_s f_{sd}$$
$$= (17{,}440 \times 504 + 0.85 \times 230{,}048 \times 20 + 2512 \times 435) \times 10^{-3}$$
$$= 13{,}792\,\text{kN}$$

$$\delta = \frac{A_a f_{yd}}{N_{pl,Rd}} = \frac{17{,}440 \times 504 \times 10^{-3}}{13{,}792} = 0.64$$

which is within the permitted range, $0.2 \leq \delta \leq 0.9$.

### Long-term effects

The creep coefficient is 2.64 (refer to design example 1, Section "Steel-reinforced concrete column with normal-strength material").

### Elastic modulus of concrete considering long-term effects

The modulus of elasticity of concrete $E_{c,eff}$ due to long-term effects is 14.22 GPa (refer to Section "Steel-reinforced concrete column with normal-strength material").

### Effective flexural stiffness of cross-section

The effective elastic flexural stiffness (refer to Section "Steel-reinforced concrete column with normal-strength material") is:

$$(EI)_{\text{eff,y}} = 1.25 \times 10^{11} \text{ kN mm}^2$$
$$(EI)_{\text{eff,z}} = 8.04 \times 10^{10} \text{ kN mm}^2$$

### Elastic critical normal force

Refer to Section "Steel-reinforced concrete column with normal-strength material":

$$N_{\text{cry}} = 77{,}000 \text{ kN}$$
$$N_{\text{crz}} = 49{,}600 \text{ kN}$$

The characteristic value of the plastic resistance to the axial load is:

$$
\begin{aligned}
N_{\text{pl,Rk}} &= A_a f_y + 0.85 A_c f_{ck} + A_s f_{sk} \\
&= (17{,}440 \times 504 + 0.85 \times 230{,}048 \times 30 + 2512 \times 500) \times 10^3 \\
&= 15{,}912 \text{ kN}
\end{aligned}
$$

### Relative slenderness ratio

$$\overline{\lambda}_y = \sqrt{\frac{N_{\text{pl,Rk}}}{N_{\text{cry}}}} = \sqrt{\frac{15{,}912}{77{,}000}} = 0.45$$

$$\overline{\lambda}_z = \sqrt{\frac{N_{\text{pl,Rk}}}{N_{\text{crz}}}} = \sqrt{\frac{15{,}912}{49{,}600}} = 0.57$$

### Buckling reduction factor

Buckling curve $b$ is applicable to axis $y$-$y$, and buckling curve $c$ is applicable to axis $z$-$z$ in accordance with EN 1994-1-1. The imperfection factor is taken as 0.34 for curve $b$ and 0.49 for curve $c$. So:

$$
\begin{aligned}
\Phi_y &= 0.5\left(1 + \alpha\left(\overline{\lambda}_y - 0.2\right) + \overline{\lambda}_y^2\right) \\
&= 0.5 \times [1 + 0.34 \times (0.45 - 0.2) + 0.45^2] = 0.65
\end{aligned}
$$

$$\Phi_z = 0.5\left(1 + \alpha\left(\overline{\lambda_z} - 0.2\right) + \overline{\lambda_z^2}\right)$$
$$= 0.5 \times [1 + 0.49 \times (0.57 - 0.2) + 0.57^2] = 0.75$$

The reduction factor for column buckling is:

$$\chi_y = \frac{1}{\Phi_y + \sqrt{\Phi_y^2 - \overline{\lambda_y^2}}} = \frac{1}{0.65 + \sqrt{0.65^2 - 0.45^2}} = 0.90$$

$$\chi_z = \frac{1}{\Phi_z + \sqrt{\Phi_z^2 - \overline{\lambda_z^2}}} = \frac{1}{0.75 + \sqrt{0.75^2 - 0.57^2}} = 0.81$$

**Buckling resistance**

The minor axis is the more critical, so

$$N_{b,Rd} = \min(\chi_y; \chi_z)N_{pl,Rd}$$
$$= 0.81 \times 13,793 = 11,172\,\text{kN} > N_{Ed} = 8000\,\text{kN}$$

The buckling resistance of the SRC column is adequate.

Compared to the SRC column with steel grade S355, the buckling resistance ratio is:

$$\frac{N_{b,Rd,S550}}{N_{b,Rd,S355}} = \frac{11,172}{9292} = 1.20$$

The buckling resistance of the SRC column with high-strength steel S550 is increased by 20% compared to the resistance of the column with S355 steel.

## Steel-reinforced concrete column with high-strength concrete and steel

Steel grade S550 and concrete class C90/105 are used in this design example. Other design data are the same as in Section "Steel-reinforced concrete column with normal-strength material," such as loading; column length; dimensions of the SRC column cross-section, steel section, and reinforcement; and so on.

### Design strengths and modulus

According to Tables 2.12 and 2.13 in the companion book, for the strain compatibility between steel and concrete, the two materials (concrete class C90/105 and steel grade S550) are strain compatible, so the steel can reach its full strength when the composite concrete section reaches its ultimate strength without considering the confinement effect from the lateral hoops and steel section.

The effective compressive strength and elastic modulus of concrete C90/105 are:

$$f_{ck} = 72 \text{ N/mm}^2; f_{cd} = 48 \text{ N/mm}^2; E_{cm} = 41.1 \text{ GPa};$$

The design strength of steel is:

$$f_y = 550 \text{ N/mm}^2; f_{yd} = 550 \text{ N/mm}^2;$$

### Cross-sectional areas and second moments of area

The cross-sectional area and second moment area of the steel, reinforcement, and concrete are the same as the design example in Section "Steel-reinforced concrete column with normal-strength material."

$A_a = 17,440 \text{ mm}^2, A_s = 2512 \text{ mm}^2, A_c = 230,048 \text{ mm}^2$
$I_{ay} = 328.1 \times 10^6 \text{ mm}^4, I_{az} = 107 \times 10^6 \text{ mm}^4$
$I_{sy} = 73.56 \times 10^6 \text{ mm}^4, I_{sz} = 73.56 \times 10^6 \text{ mm}^4$
$I_{cy} = 4805 \times 10^6 \text{ mm}^4, I_{cz} = 5026 \times 10^6 \text{ mm}^4$

### Check the steel contribution factor

The design plastic resistance of the composite cross-section in compression is:

$$\begin{aligned} N_{pl,Rd} &= A_a f_{yd} + 0.85 A_c f_{cd} + A_s f_{sd} \\ &= (17,440 \times 550 + 0.85 \times 230,048 \times 48 + 2512 \times 435) \times 10^{-3} \\ &= 20,070 \text{kN} \end{aligned}$$

$$\delta = \frac{A_a f_{yd}}{N_{pl,Rd}} = \frac{17,440 \times 550 \times 10^{-3}}{11,195} = 0.85$$

which is within the permitted range, $0.2 \leq \delta \leq 0.9$.

### Long-term effects

The creep coefficient is 1.30 (refer to design example 2, Section "Steel-reinforced concrete column with high-strength concrete").

### Elastic modulus of concrete considering long-term effects

The modulus of elasticity of concrete $E_{c,eff}$ due to long-term effects is 24.9 GPa (refer to Section "Steel-reinforced concrete column with high-strength concrete").

### Effective flexural stiffness of cross-section

The effective elastic flexural stiffness (refer to Section "Steel-reinforced concrete column with high-strength concrete") is:

$$(EI)_{eff,y} = 1.56 \times 10^{11} \text{ kNmm}^2$$
$$(EI)_{eff,z} = 1.13 \times 10^{11} \text{ kNmm}^2$$

### Elastic critical normal force

Refer to Section "Steel-reinforced concrete column with high-strength concrete":

$$N_{cry} = 96,100 \text{ kN}$$
$$N_{crz} = 69,500 \text{ kN}$$

The characteristic value of the plastic resistance to the axial load is:

$$
\begin{aligned}
N_{pl,Rk} &= A_a f_y + 0.85 A_c f_{ck} + A_s f_{sk} \\
&= (17,440 \times 550 + 0.85 \times 230,048 \times 72 + 2512 \times 500) \times 10^3 \\
&= 24,927 \text{ kN}
\end{aligned}
$$

### Relative slenderness ratio

$$\overline{\lambda}_y = \sqrt{\frac{N_{pl,Rk}}{N_{cry}}} = \sqrt{\frac{24,927}{96,100}} = 0.51$$

$$\overline{\lambda}_z = \sqrt{\frac{N_{pl,Rk}}{N_{crz}}} = \sqrt{\frac{24,927}{69,500}} = 0.60$$

### Buckling reduction factor

Buckling curve $b$ is applicable to axis $y$-$y$, and buckling curve $c$ is applicable to axis $z$-$z$ in accordance with EN 1994-1-1. The imperfection factor is taken as 0.34 for curve $b$ and 0.49 for curve $c$. So:

$$\Phi_y = 0.5\left(1+\alpha\left(\overline{\lambda_y}-0.2\right)+\overline{\lambda_y^2}\right)$$
$$= 0.5\times[1+0.34\times(0.51-0.2)+0.51^2]=0.68$$

$$\Phi_z = 0.5\left(1+\alpha\left(\overline{\lambda_z}-0.2\right)+\overline{\lambda_z^2}\right)$$
$$= 0.5\times[1+0.49\times(0.60-0.2)+0.60^2]=0.78$$

The reduction factor for column buckling is:

$$\chi_y = \frac{1}{\Phi_y+\sqrt{\Phi_y^2-\overline{\lambda_y^2}}} = \frac{1}{0.68+\sqrt{0.68^2-0.51^2}} = 0.88$$

$$\chi_z = \frac{1}{\Phi_z+\sqrt{\Phi_z^2-\overline{\lambda_z^2}}} = \frac{1}{0.78+\sqrt{0.78^2-0.60^2}} = 0.79$$

### Buckling resistance

The minor axis is the more critical, so

$$N_{b,Rd} = \min(\chi_y; \chi_z)N_{pl,Rd}$$
$$= 0.79\times20{,}070 = 15{,}855\,\text{kN} > N_{Ed} = 8000\,\text{kN}$$

The buckling resistance of the SRC column is adequate.

Compared to the SRC column with normal-strength material S355 and C30/37, the buckling resistance ratio is:

$$\frac{N_{b,Rd,H}}{N_{b,Rd,N}} = \frac{15{,}855}{9292} = 1.71$$

The buckling resistance of the SRC column with high-strength steel S550 and high-strength concrete C90/105 is increased by 71% compared to the resistance of the column with normal-strength steel S355 and normal-strength concrete C30/37.

## Alternative design

Alternatively, the column size can be reduced when high-strength steel and concrete materials are used, but the buckling resistance is almost the same as in design example 1 in Section "Steel-reinforced concrete column with normal-strength material."

## Design data

| | |
|---|---|
| Structural steel | Grade S550 |
| Concrete | C90/105 |

**Properties of cross-section**

| | |
|---|---|
| Concrete depth | $h_c = 400$ mm |
| Concrete width | $b_c = 400$ mm |
| Concrete cover | $c = 30$ mm |
| Cover | $c_y = 71.9$ mm |
| Cover | $c_z = 69.9$ mm |

**Section properties of steel section**
254 × 254 UC 89

| | |
|---|---|
| Depth | $h = 260.3$ mm |
| Width | $b = 256.3$ mm |
| Flange thickness | $t_f = 27.3$ mm |
| Web thickness | $t_w = 10.3$ mm |
| Fillet | $r = 12.7$ mm |
| Section area | $A_a = 113.3$ cm² |
| Second moment of area/y | $I_{ay} = 14,270$ cm⁴ |
| Second moment of area/z | $I_{az} = 4857$ cm⁴ |
| Plastic section modulus/y | $W_{pl,a,y} = 1224$ cm³ |
| Plastic section modulus/z | $W_{pl,a,z} = 575$ cm³ |

Other data are the same as in design example 1 in Section "Steel-reinforced concrete column with normal-strength material."

## Design strengths and modulus

According to Tables 2.12 and 2.13 in the companion book, for the strain compatibility between steel and concrete, the two materials (concrete class

C90/105 and steel grade S550) are strain compatible, so the steel can reach its full strength when the composite concrete section reaches its ultimate strength without considering the confinement effect from the lateral hoops and steel section.

The effective compressive strength and elastic modulus of concrete C90/105 are:

$$f_{ck} = 72 \text{ N/mm}^2; f_{cd} = 48 \text{ N/mm}^2; E_{cm} = 41.1 \text{ GPa}$$

The design strength of steel is:

$$f_y = 550 \text{ N/mm}^2; f_{yd} = 550 \text{ N/mm}^2$$

### Cross-sectional area and second moments of area

The cross-sectional area and second moment area of the steel, reinforcement, and concrete are:

$$A_a = 11{,}330 \text{ mm}^2, A_s = 2512 \text{ mm}^2, A_c = 146{,}158 \text{ mm}^2$$
$$I_{ay} = 142.7 \times 10^6 \text{ mm}^4, I_{az} = 48.57 \times 10^6 \text{ mm}^4$$
$$I_{sy} = 42.39 \times 10^6 \text{ mm}^4, I_{sz} = 42.39 \times 10^6 \text{ mm}^4$$
$$I_{cy} = 1948 \times 10^6 \text{ mm}^4, I_{cz} = 2042 \times 10^6 \text{ mm}^4$$

### Check the reinforcement ratio

$$\rho_s = \frac{A_s}{A_c} = \frac{2512}{146{,}158} = 1.7\%$$

The reinforcement ratio is within the range 0.3%–6%.

### Check the local buckling

The concrete cover to the flange of the steel section: $c = 69.9$ mm $>$ maximum (40 mm; $b_f/6$).

Thus, the effect of local buckling is neglected for the SRC column.

### Check the steel contribution factor

The design plastic resistance of the composite cross-section in compression is:

$$N_{\text{pl,Rd}} = A_a f_{yd} + 0.85 A_c f_{cd} + A_s f_{sd}$$
$$= (11{,}330 \times 550 + 0.85 \times 146{,}158 \times 48 + 2512 \times 435) \times 10^{-3}$$
$$= 13{,}287\,\text{kN}$$

$$\delta = \frac{A_a f_{yd}}{N_{\text{pl,Rd}}} = \frac{11{,}330 \times 550 \times 10^{-3}}{13{,}287} = 0.47 < 0.9$$

### Long-term effects

The notional size of the cross-section is:

$$h_0 = 2A_c/u = 183 \text{ mm}$$

Coefficient:

$$\alpha_1 = \left(\frac{35}{f_{cm}}\right)^{0.7} = \left(\frac{35}{72+8}\right)^{0.7} = 0.56$$

$$\alpha_2 = \left(\frac{35}{f_{cm}}\right)^{0.2} = \left(\frac{35}{80}\right)^{0.2} = 0.85$$

$$\alpha_3 = \left(\frac{35}{f_{cm}}\right)^{0.5} = \left(\frac{35}{80}\right)^{0.5} = 0.66$$

Factor:

$$\varphi_{\text{RH}} = \left[1 + \frac{1 - RH/100}{0.1\sqrt[3]{h_0}} \alpha_1\right]\alpha_2$$
$$= \left[1 + \frac{1 - 50/100}{0.1\sqrt[3]{183}} \times 0.56\right] \times 0.85 = 1.27$$

$$\beta(f_{cm}) = \frac{16.8}{\sqrt{f_{cm}}} = \frac{16.8}{\sqrt{80}} = 1.88$$

$$\beta(t_0) = \frac{1}{(0.1 + t_0^{0.20})} = \frac{1}{(0.1 + 14^{0.20})} = 0.56$$

$$\varphi_0 = \varphi_{\mathrm{RH}}\beta(f_{\mathrm{cm}})\beta(t_0) = 1.27 \times 1.88 \times 0.56 = 1.34$$

Factor:

$$\beta_{\mathrm{H}} = 1.5\left[1+(0.012RH)^{18}\right]h_0 + 250\alpha_3$$
$$= 1.5\left[1+(0.012\times50)^{18}\right]\times183 + 250\times0.66 = 440$$

$$\beta(t,t_0) = \left[\frac{t-t_0}{(\beta_{\mathrm{H}}+t-t_0)}\right]^{0.3} = \left[\frac{\infty-14}{(440+\infty-14)}\right]^{0.3} = 1.0$$

The creep coefficient is:

$$\varphi_t = \varphi_0\beta_{\mathrm{c}}(t,t_0) = 1.34 \times 1.0 = 1.34$$

### Elastic modulus of concrete considering long-term effects

$$E_{\mathrm{c,eff}} = \frac{E_{\mathrm{cm}}}{1+(N_{\mathrm{G,Ed}}/N_{\mathrm{Ed}})\varphi_t} = \frac{41.1}{1+(4000/8000)\times1.34} = 24.6\,\mathrm{kN/mm}^2$$

### Effective flexural stiffness of cross-section

$$(EI)_{\mathrm{eff,y}} = E_{\mathrm{a}}I_{\mathrm{ay}} + 0.6E_{\mathrm{c,eff}}I_{\mathrm{cy}} + E_{\mathrm{s}}I_{\mathrm{sy}} = 6.73\times10^{10}\,\mathrm{kN\,mm}^2$$

$$(EI)_{\mathrm{eff,z}} = E_{\mathrm{a}}I_{\mathrm{az}} + 0.6E_{\mathrm{c,eff}}I_{\mathrm{cz}} + E_{\mathrm{s}}I_{\mathrm{sz}} = 4.89\times10^{10}\,\mathrm{kN\,mm}^2$$

### Elastic critical normal force

$$N_{\mathrm{cry}} = \frac{\pi^2(EI)_{\mathrm{eff,y}}}{L_{\mathrm{y}}^2} = \frac{\pi^2\times6.73\times10^{10}}{4^2\times10^6} = 41{,}500\,\mathrm{kN}$$

$$N_{\mathrm{crz}} = \frac{\pi^2(EI)_{\mathrm{eff,z}}}{L_{\mathrm{z}}^2} = \frac{\pi^2\times4.89\times10^{10}}{4^2\times10^6} = 30{,}200\,\mathrm{kN}$$

The characteristic value of the plastic resistance to the axial load is:

$$N_{\text{pl,Rk}} = A_a f_y + 0.85 A_c f_{ck} + A_s f_{sk} = 16,432\,\text{kN}$$

### Relative slenderness ratio

$$\overline{\lambda_y} = \sqrt{\frac{N_{\text{pl,Rk}}}{N_{cry}}} = \sqrt{\frac{16,432}{41,500}} = 0.63$$

$$\overline{\lambda_z} = \sqrt{\frac{N_{\text{pl,Rk}}}{N_{crz}}} = \sqrt{\frac{16,432}{30,200}} = 0.74$$

### Buckling reduction factor

Buckling curve $b$ is applicable to axis $y$-$y$, and buckling curve $c$ is applicable to axis $z$-$z$ in accordance with EN 1994-1-1. The imperfection factor is taken as 0.34 for curve $b$ and 0.49 for curve $c$. So:

$$\Phi_y = 0.5\left(1 + \alpha\left(\overline{\lambda_y} - 0.2\right) + \overline{\lambda_y^2}\right)$$
$$= 0.5 \times [1 + 0.34 \times (0.63 - 0.2) + 0.63^2] = 0.77$$

$$\Phi_z = 0.5\left(1 + \alpha\left(\overline{\lambda_z} - 0.2\right) + \overline{\lambda_z^2}\right)$$
$$= 0.5 \times [1 + 0.49 \times (0.74 - 0.2) + 0.74^2] = 0.90$$

The reduction factor for column buckling is:

$$\chi_y = \frac{1}{\Phi_y + \sqrt{\Phi_y^2 - \overline{\lambda_y^2}}} = \frac{1}{0.77 + \sqrt{0.77^2 - 0.63^2}} = 0.82$$

$$\chi_z = \frac{1}{\Phi_z + \sqrt{\Phi_z^2 - \overline{\lambda_z^2}}} = \frac{1}{0.90 + \sqrt{0.90^2 - 0.74^2}} = 0.70$$

### Buckling resistance

The minor axis is the more critical, so

$$N_{b,Rd} = \min(\chi_y; \chi_z) N_{pl,Rd}$$
$$= 0.70 \times 13,287 = 9300\,\text{kN} > N_{Ed} = 8000\,\text{kN}$$

The buckling resistance of the SRC column is adequate.

Compared to the SRC column with normal-strength materials S355 and C30/37, the buckling resistance ratio is:

$$\frac{N_{b,Rd,H}}{N_{b,Rd,N}} = \frac{9300}{9292} \approx 1.0$$

The buckling resistance is almost the same as that of the SRC column with normal-strength materials S355 and C30/37.

The cross-section area ratio of the SRC column is:

$$\frac{A_H}{A_N} = \frac{400 \times 400}{500 \times 500} = 0.64$$

The cross-section area of the SRC column with high-strength materials S550 and C90/105 is reduced by 36% compared to the SRC column with S355 steel and C30/37 concrete. Similarly, the amount of the steel section is also reduced by 36% compared to the SRC column with normal-strength material.

## STEEL-REINFORCED CONCRETE COLUMN SUBJECTED TO COMBINED COMPRESSION AND BENDING

Determine the resistance of SRC columns subjected to compression and bending about the major axis.

### Steel-reinforced concrete column with normal-strength material

#### Design data

| | |
|---|---|
| Design axial force | $N_{Ed} = 9000$ kN |
| Permanent load | $N_{G,Ed} = 4000$ kN |
| Design moment | $M_{b,y} = 300$ kNm |
| | $M_{t,y} = 200$ kNm |

*Continued*

| Column length | $L = 4.0$ m |
|---|---|
| Effective length | $L_{eff} = 4.0$ m |
| Structural steel | Grade S355, $f_y = 355$ N/mm² |
| Concrete | C50/60, $f_{ck} = 50$ N/mm² |
| Reinforcement | $f_{sk} = 500$ N/mm² |

**Properties of cross-section:**

| Concrete depth | $h_c = 500$ mm |
|---|---|
| Concrete width | $b_c = 500$ mm |
| Concrete cover | $c = 30$ mm |
| Cover | $c_y = 96.3$ mm |
| Cover | $c_z = 92.8$ mm |

**Section properties of steel section**
305 × 305 UC 118

| Depth | $h = 314.5$ mm |
|---|---|
| Width | $b = 307.4$ mm |
| Flange thickness | $t_f = 18.7$ mm |
| Web thickness | $t_w = 12$ mm |
| Fillet | $r = 15.2$ mm |
| Section area | $A_a = 150.2$ cm² |
| Second moment of area/y | $I_{ay} = 272{,}670$ cm⁴ |
| Second moment of area/z | $I_{az} = 9059$ cm⁴ |
| Plastic section modulus/y | $W_{pl,a,y} = 1958$ cm³ |
| Plastic section modulus/z | $W_{pl,a,z} = 895$ cm³ |

**Reinforcement**

| Longitudinal reinforcement | number $n = 8$, diameter $d_{l,s} = 20$ mm |
|---|---|
| Transverse reinforcement | diameter $d_{t,s} = 10$ mm, spacing $s = 200$ mm |

## Design strengths and modulus

According to Tables 2.12 and 2.13 in the companion book, for the strain compatibility between steel and concrete, the two materials (concrete class C50/60 and steel grade S355) are strain compatible, so the steel can reach its full strength when the composite concrete section reaches its ultimate strength without considering the confinement effect from the lateral hoops and steel section.

Then, the design strengths of the steel, concrete, and reinforcement are:

$$f_{yd} = \frac{f_y}{\gamma_M} = \frac{355}{1.0} = 355\,\text{N/mm}^2$$

$$f_{cd} = \frac{f_{ck}}{\gamma_C} = \frac{50}{1.5} = 33.3\,\text{N/mm}^2$$

$$f_{sd} = \frac{f_{sk}}{\gamma_S} = \frac{500}{1.15} = 435\,\text{N/mm}^2$$

$$E_{cm} = 37\,\text{Gpa}$$

$$E_a = 210\,\text{Gpa}$$

## Cross-sectional areas and second moments of area

$A_a = 15{,}020\,\text{mm}^2$, $A_s = 2512\,\text{mm}^2$, $A_c = 232{,}468\,\text{mm}^2$
$I_{ay} = 276.7 \times 10^6\,\text{mm}^4$, $I_{az} = 90.6 \times 10^6\,\text{mm}^4$
$I_{sy} = 75.36 \times 10^6\,\text{mm}^4$, $I_{sz} = 75.36 \times 10^6\,\text{mm}^4$
$I_{cy} = 4856 \times 10^6\,\text{mm}^4$, $I_{cz} = 5042 \times 10^6\,\text{mm}^4$

## Check the reinforcement ratio

$$\rho_s = \frac{A_s}{A_c} = \frac{2512}{232{,}468} = 1.1\%$$

The reinforcement ratio is within the range 0.3%–6%.

## Check the local buckling

The concrete cover to the flange of the steel section: $c = 92.8$ mm > maximum (40 mm; $b_f/6$).

Thus, the effect of local buckling is neglected for the SRC column.

## Check the steel contribution factor

The design plastic resistance of the composite cross-section in compression is:

$$N_{\text{pl,Rd}} = A_a f_{yd} + 0.85 A_c f_{cd} + A_s f_{sd}$$
$$= (15{,}020 \times 355 + 0.85 \times 232{,}468 \times 33.3 + 2512 \times 435) \times 10^{-3}$$
$$= 13{,}011\,\text{kN}$$

$$\delta = \frac{A_a f_{yd}}{N_{\text{pl,Rd}}} = \frac{15{,}020 \times 355 \times 10^{-3}}{13{,}011} = 0.41 < 0.9$$

### Long-term effects

The age of concrete at loading $t_0$ is assumed to be 28 days. The age of concrete at the moment considered $t$ is taken as infinity. The relative humidity $RH$ is taken as 50%.

The notional size of the cross-section is:

$$h_0 = \frac{2 A_c}{u} = \frac{2 \times 232{,}468}{4 \times 500} = 232\,\text{mm}$$

Coefficient:

$$\alpha_1 = \left( \frac{35}{f_{cm}} \right)^{0.7} = \left( \frac{35}{58} \right)^{0.7} = 0.70$$

$$\alpha_2 = \left( \frac{35}{f_{cm}} \right)^{0.2} = \left( \frac{35}{58} \right)^{0.2} = 0.90$$

$$\alpha_3 = \left( \frac{35}{f_{cm}} \right)^{0.5} = \left( \frac{35}{58} \right)^{0.5} = 0.78$$

Factor:

$$\varphi_{\text{RH}} = \left[ 1 + \frac{1 - RH/100}{0.1 \sqrt[3]{h_0}} \alpha_1 \right] \alpha_2 = \left[ 1 + \frac{1 - 50/100}{0.1 \sqrt[3]{232}} \times 0.70 \right] \times 0.90 = 1.41$$

$$\beta(f_{cm}) = \frac{16.8}{\sqrt{f_{cm}}} = \frac{16.8}{\sqrt{58}} = 2.21$$

$$\beta(t_0) = \frac{1}{(0.1 + t_0^{0.20})} = \frac{1}{(0.1 + 28^{0.20})} = 0.49$$

$$\varphi_0 = \varphi_{\text{RH}}\beta(f_{\text{cm}})\beta(t_0) = 1.41 \times 2.21 \times 0.49 = 1.53$$

Factor:

$$\beta_{\text{H}} = 1.5\,[1 + (0.012\text{RH})^{18}]\,h_0 + 250\alpha_3$$
$$= 1.5\,[1 + (0.012 \times 50)^{18}] \times 232 + 250 \times 0.78 = 543$$

$$\beta(t,t_0) = \left[\frac{t - t_0}{(\beta_{\text{H}} + t - t_0)}\right]^{0.3} = \left[\frac{\infty - 28}{(543 + \infty - 28)}\right]^{0.3} = 1.0$$

The creep coefficient is:

$$\varphi_t = \varphi_0\beta_c(t,t_0) = 1.53 \times 1.0 = 1.53$$

### Elastic modulus of concrete considering long-term effects

Long-term effects due to creep and shrinkage should be considered in determining the effective elastic flexural stiffness. The modulus of elasticity of concrete $E_{\text{cm}}$ is reduced to the value $E_{\text{c,eff}}$:

$$E_{\text{c,eff}} = \frac{E_{\text{cm}}}{1 + (N_{\text{G,Ed}}/N_{\text{Ed}})\varphi_t} = \frac{37}{1 + (4000/9000) \times 1.53} = 22.2\,\text{kN/mm}^2$$

### Effective flexural stiffness of cross-section

The effective elastic flexural stiffness taking account of the long-term effects is:

$$(EI)_{\text{eff,y}} = E_a I_{\text{ay}} + 0.6E_{\text{c,eff}} I_{\text{cy}} + E_s I_{\text{sy}}$$
$$= 210 \times 276.7 \times 10^6 + 0.6 \times 22.2 \times 4856 \times 10^6 + 210 \times 75.36 \times 10^6$$
$$= 1.38 \times 10^{11}\,\text{kN}\,\text{mm}^2$$

$$(EI)_{\text{eff,z}} = E_a I_{az} + 0.6 E_{c,\text{eff}} I_{cz} + E_s I_{sz}$$
$$= 210 \times 90.6 \times 10^6 + 0.6 \times 22.2 \times 5042 \times 10^6 + 210 \times 75.36 \times 10^6$$
$$= 1.01 \times 10^{11}\,\text{kN}\,\text{mm}^2$$

### Elastic critical normal force

$$N_{\text{cry}} = \frac{\pi^2 (EI)_{\text{eff,y}}}{L_y^2} = \frac{\pi^2 \times 1.38 \times 10^{11}}{4^2 \times 10^6} = 85,100\,\text{kN}$$

$$N_{\text{crz}} = \frac{\pi^2 (EI)_{\text{eff,z}}}{L_z^2} = \frac{\pi^2 \times 1.01 \times 10^{11}}{4^2 \times 10^6} = 62,500\,\text{kN}$$

The characteristic value of the plastic resistance to the axial load is:

$$N_{\text{pl,Rk}} = A_a f_y + 0.85 A_c f_{ck} + A_s f_{sk}$$
$$= (15,020 \times 355 + 0.85 \times 232,468 \times 50 + 2512 \times 500) \times 10^3$$
$$= 16,468\,\text{kN}$$

### Relative slenderness ratio

$$\overline{\lambda}_y = \sqrt{\frac{N_{\text{pl,Rk}}}{N_{\text{cry}}}} = \sqrt{\frac{16,468}{85,100}} = 0.44$$

$$\overline{\lambda}_z = \sqrt{\frac{N_{\text{pl,Rk}}}{N_{\text{crz}}}} = \sqrt{\frac{16,468}{62,500}} = 0.51$$

### Buckling reduction factor

Buckling curve *b* is applicable to axis *y-y*, and buckling curve *c* is applicable to axis *z-z* in accordance with EN 1994-1-1. The imperfection factor is taken as 0.34 for curve *b* and 0.49 for curve *c*. According to EN 1993-1-1, the factor is:

$$\Phi_y = 0.5\left(1 + \alpha\left(\overline{\lambda}_y - 0.2\right) + \overline{\lambda}_y^2\right)$$
$$= 0.5 \times [1 + 0.34 \times (0.44 - 0.2) + 0.44^2] = 0.64$$

$$\Phi_z = 0.5\left(1 + \alpha\left(\overline{\lambda}_z - 0.2\right) + \overline{\lambda}_z^2\right)$$
$$= 0.5 \times [1 + 0.49 \times (0.51 - 0.2) + 0.51^2] = 0.71$$

The reduction factor for column buckling is:

$$\chi_y = \frac{1}{\Phi_y + \sqrt{\Phi_y^2 - \overline{\lambda}_y^2}} = \frac{1}{0.64 + \sqrt{0.64^2 - 0.44^2}} = 0.91$$

$$\chi_z = \frac{1}{\Phi_z + \sqrt{\Phi_z^2 - \overline{\lambda}_z^2}} = \frac{1}{0.71 + \sqrt{0.71^2 - 0.51^2}} = 0.84$$

## Buckling resistance

The minor axis is the more critical, so

$$N_{b,Rd} = \min(\chi_y; \chi_z)N_{pl,Rd}$$
$$= 0.84 \times 13{,}011 = 10{,}929\,\text{kN} > N_{Ed} = 9000\,\text{kN}$$

The buckling resistance of the SRC column is adequate.

## Interaction curve

The polygonal interaction diagram for major-axis bending is calculated, using the notation shown in Figure 2.

### Point A $(0, N_{pl,Rd})$

The full cross-section is under compression without the bending moment.

$$M_A = 0$$
$$N_A = N_{pl,Rd} = 13{,}011\ \text{kN}$$

### Point B $(M_{pl,Rd}, 0)$

Assuming the neutral axis lies in the web of the steel section ($h_n \le h/2 - t_f$), the 2 reinforcement bar lies within the region $2h_n$, $A_{sn} = 628$ mm², so,

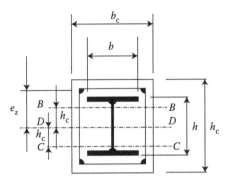

*Figure 2* Plastic neutral axes for encased I-section.

$$h_n = \frac{A_c 0.85 f_{cd} - A_{sn}(2f_{sd} - 0.85 f_{cd})}{2b_c 0.85 f_{cd} + 2t_w(2f_{yd} - 0.85 f_{cd})}$$

$$= \frac{232,468 \times 0.85 \times 33.3 - 628 \times (2 \times 435 - 0.85 \times 33.3)}{2 \times 500 \times 0.85 \times 33.3 + 2 \times 12 \times (2 \times 355 - 0.85 \times 33.3)} = 136\,\text{mm}$$

Hence,

$$h_n = 136\,\text{mm} < \frac{h}{2} - t_f = 138.6\,\text{mm}$$

The assumption for the plastic neutral axis is verified. The neutral axis lies in the web of the steel section.

The plastic section moduli for the steel section, reinforcement, and concrete are:

$$W_{pa} = 1.958 \times 10^6\,\text{mm}^3$$

$$W_{ps} = \sum_1^6 A_{si}[e_i] = 1884 \times 200 = 0.3768 \times 10^6\,\text{mm}^3$$

$$W_{pc} = \frac{b_c h_c^2}{4} - W_{pa} - W_{ps} = \frac{500 \times 500^2}{4} - 1.958 \times 10^6 - 0.3768 \times 10^6$$

$$= 28.915 \times 10^6\,\text{mm}^3$$

The plastic section moduli for the region of depth $2h_n$ are:

$$W_{psn} = 0\,\text{mm}^3$$

$$W_{pan} = t_w h_n^2 = 12 \times 136^2 = 0.222 \times 10^6\,\text{mm}^3$$

$$W_{pcn} = b_c h_n^2 - W_{pan} - W_{psn} = 500 \times 136^2 - 0.222 \times 10^6 - 0$$
$$= 9.026 \times 10^6\,\text{mm}^3$$

The bending resistance at point B is determined from:

$$M_{pl,Rd} = (W_{pa} - W_{pa,n})f_{yd} + (W_{ps} - W_{ps,n})f_{sd} + 0.5(W_{pc} - W_{pc,n})\alpha_c f_{cd}$$
$$= (1.958 - 0.222) \times 355 + (0.3768 - 0) \times 435$$
$$+ 0.5 \times (28.915 - 9.026) \times 0.85 \times 33.3 = 1062\,\text{kNm}$$

## Point C ($M_{pl,Rd}$, $N_{pm,Rd}$)

The axial force is equal to the full cross-section compression resistance of concrete. The value is determined from:

$$N_{pm,Rd} = 0.85 A_c f_{cd} = 0.85 \times 232,468 \times 33.3 \times 10^{-3} = 6850\,\text{kN}$$

## Point D ($M_{max,Rd}$, $0.5N_{pm,Rd}$)

The maximum moment resistance is determined from:

$$M_{max,Rd} = f_{yd}W_{pa} + 0.5 \times 0.85 f_{cd}W_{pc} + f_{sd}W_{ps}$$
$$= 355 \times 1.958 + 0.5 \times 0.85 \times 33.3 \times 28.915 + 435 \times 0.3768$$
$$= 1268\,\text{kNm}$$

The relative information for plotting the interaction curve is shown in Table 1. Then, the interaction curve is plotted as shown in Figure 3.

### Check the resistance of steel-reinforced concrete column in combined compression and uniaxial bending

The effective flexural stiffness considering second-order effects is determined from:

*Table I* The resistance for interaction curve

| Point | Resistance to bending (kNm) | Resistance to compression (kN) |
|-------|------------------------------|--------------------------------|
| A | 0 | 13,011 |
| B | 1062 | 0 |
| C | 1062 | 6850 |
| D | 1268 | 3425 |

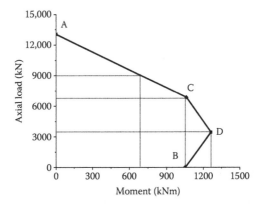

*Figure 3* Interaction curve for major axis.

$$(EI)_{\text{eff,II,y}} = K_o(E_a I_{\text{ay}} + K_{\text{e,II}} E_{\text{c,eff}} I_{\text{cy}} + E_s I_{\text{sy}})$$
$$= 0.9 \times (210 \times 276.7 \times 10^6 + 0.5 \times 22.2$$
$$\times 4856 \times 10^6 + 210 \times 75.36 \times 10^6) = 1.15 \times 10^{11} \text{ kNmm}^2$$

Hence, the elastic critical force is:

$$N_{\text{cr,y,eff}} = \frac{\pi^2 (EI)_{\text{eff,II,y}}}{L_y^2} = \frac{\pi^2 \times 1.15 \times 10^{11}}{4^2 \times 10^6} = 70,600 \text{ kN}$$

The result is less than $10 N_{\text{Ed}}$ for major axis, so the second-order effects must be considered for the moment from first-order analysis and the moment from imperfection.

The member imperfection for the major axis according to EN 1994-1-1 is:

$$e_{0,y} = L/200 = 20 \text{ mm}$$

For the major axis, the midlength bending moments due to $N_{Ed}$ and imperfection are calculated by:

$$N_{Ed}e_{0,y} = 9000 \times 0.02 = 180 \, kNm$$

According to EN 1994-1-1, the factor $\beta$ is equal to 1.0 for the bending moment from member imperfection. Then, the amplification factor is:

$$k_{imp,y} = \frac{\beta}{1 - N_{Ed}/N_{cr,y,eff}} = \frac{1.0}{1 - 9000/70,600} = 1.15$$

For the first-order bending moment, $M_{y,top} = 200 \, kNm$, $M_{y,bot} = 300 \, kNm$, so the ratio of the end moment is:

$$r = 200/300 = 0.667$$

Then, the factor $\beta$ is:

$$\beta = \max \, (0.66 + 0.44 \, r; \, 0.44) = 0.95$$

thus, the amplification factor is:

$$k_y = \frac{\beta}{1 - N_{Ed}/N_{cr,y,eff}} = \frac{0.95}{1 - 9000/70600} = 1.09$$

Hence, the design moment considering second-order effects is:

$$M_{y,Ed} = k_y M_{y,Ed,top} + k_{imp,y} N_{Ed} e_{0,z} = 1.09 \times 300 + 1.15 \times 200$$
$$= 557 \, kNm$$

For $N_{Ed} > N_{pm,Rd}$, the factor is determined from:

$$\mu_d = \frac{N_{pl,Rd} - N_{Ed}}{N_{pl,Rd} - N_{pm,Rd}} = \frac{13,011 - 9000}{13,011 - 6850} = 0.65$$

Thus,

$$\frac{M_{y,Ed}}{M_{pl,N,y,Rd}} = \frac{M_{y,Ed}}{\mu_d M_{pl,y,Rd}} = \frac{557}{0.65 \times 1062} = 0.81 < 0.9$$

So, the resistance of the SRC column to compression and uniaxial bending is satisfied.

## Steel-reinforced concrete column with high-strength concrete

The concrete class C90/105 is used in this design example. Other design data are same as in Section "Steel-reinforced concrete column with normal-strength material," such as loading; column length; steel strength; and dimensions of the SRC column cross-section, steel section, and reinforcement.

### Design strengths and modulus

According to Tables 2.12 and 2.13 in the companion book, for the strain compatibility between steel and concrete, the two materials (concrete class C90/105 and steel grade S355) are strain compatible, so the steel can reach its full strength when the composite concrete section reaches its ultimate strength without considering the confinement effect from the lateral hoops and steel section.

For high-strength concrete with $f_{ck} > 50 \, \text{N/mm}^2$, the effective compressive strength of concrete should be used in accordance with EC2. The effective strength is:

$$f_{ck} = 90\eta = 90 \times (1.0 - (90 - 50)/200) = 72 \, \text{N/mm}^2$$

Accordingly, the secant modulus for high-strength concrete C90/105 is:

$$E_{cm} = 22[(\eta f_{ck} + 8)/10]^{0.3} = 22[(72 + 8)/10]^{0.3} = 41.1 \, \text{GPa}$$

The design strength of concrete is:

$$f_{cd} = \frac{f_{ck}}{\gamma_C} = \frac{72}{1.5} = 48 \, \text{N/mm}^2$$

### Cross-sectional areas and second moments of area

$A_a = 15{,}020 \, \text{mm}^2$, $A_s = 2512 \, \text{mm}^2$, $A_c = 232{,}468 \, \text{mm}^2$
$I_{ay} = 276.7 \times 10^6 \, \text{mm}^4$, $I_{az} = 90.6 \times 10^6 \, \text{mm}^4$
$I_{sy} = 75.36 \times 10^6 \, \text{mm}^4$, $I_{sz} = 75.36 \times 10^6 \, \text{mm}^4$
$I_{cy} = 4856 \times 10^6 \, \text{mm}^4$, $I_{cz} = 5042 \times 10^6 \, \text{mm}^4$

### Check the steel contribution factor

The design plastic resistance of the composite cross-section in compression is:

$$N_{pl,Rd} = A_a f_{yd} + 0.85 A_c f_{cd} + A_s f_{sd}$$
$$= (15,020 \times 355 + 0.85 \times 232,468 \times 48 + 2512 \times 435) \times 10^{-3}$$
$$= 15,909 \, kN$$

$$\delta = \frac{A_a f_{yd}}{N_{pl,Rd}} = \frac{15,020 \times 355 \times 10^{-3}}{15,909} = 0.34 < 0.9$$

## Long-term effects

The age of concrete at loading $t_0$ is assumed to be 28 days. The age of concrete at the moment considered $t$ is taken as infinity. The relative humidity $RH$ is taken as 50%.

The notional size of the cross-section is:

$$h_0 = \frac{2A_c}{u} = \frac{2 \times 232,468}{4 \times 500} = 232 \, mm$$

Coefficient:

$$\alpha_1 = \left(\frac{35}{f_{cm}}\right)^{0.7} = \left(\frac{35}{80}\right)^{0.7} = 0.56$$

$$\alpha_2 = \left(\frac{35}{f_{cm}}\right)^{0.2} = \left(\frac{35}{80}\right)^{0.2} = 0.85$$

$$\alpha_3 = \left(\frac{35}{f_{cm}}\right)^{0.5} = \left(\frac{35}{80}\right)^{0.5} = 0.66$$

Factor:

$$\varphi_{RH} = \left[1 + \frac{1 - RH/100}{0.1\sqrt[3]{h_0}} \alpha_1\right] \alpha_2 = \left[1 + \frac{1 - 50/100}{0.1\sqrt[3]{232}} \times 0.56\right] \times 0.85 = 1.24$$

$$\beta(f_{cm}) = \frac{16.8}{\sqrt{f_{cm}}} = \frac{16.8}{\sqrt{80}} = 1.88$$

$$\beta(t_0) = \frac{1}{\left(0.1 + t_0^{0.20}\right)} = \frac{1}{(0.1 + 28^{0.20})} = 0.49$$

$$\varphi_0 = \varphi_{RH}\beta(f_{cm})\beta(t_0) = 1.24 \times 1.88 \times 0.49 = 1.14$$

Factor:

$$\beta_H = 1.5[1 + (0.012RH)^{18}]\, h_0 + 250\alpha_3$$
$$= 1.5[1 + (0.012 \times 50)^{18}] \times 232 + 250 \times 0.66 = 513$$

$$\beta(t,t_0) = \left[\frac{t - t_0}{(\beta_H + t - t_0)}\right]^{0.3} = \left[\frac{\infty - 28}{(513 + \infty - 28)}\right]^{0.3} = 1.0$$

The creep coefficient is:

$$\varphi_t = \varphi_0\beta_c(t,t_0) = 1.14 \times 1.0 = 1.14$$

### Elastic modulus of concrete considering long-term effects

Long-term effects due to creep and shrinkage should be considered in determining the effective elastic flexural stiffness. The modulus of elasticity of concrete $E_{cm}$ is reduced to the value $E_{c,eff}$:

$$E_{c,eff} = \frac{E_{cm}}{1 + (N_{G,Ed}/N_{Ed})\varphi_t} = \frac{41.1}{1 + (4000/9000) \times 1.14} = 27.2\,\text{kN/mm}^2$$

### Effective flexural stiffness of cross-section

The effective elastic flexural stiffness taking account of the long-term effects is:

$$(EI)_{eff,y} = E_a I_{ay} + 0.6E_{c,eff}I_{cy} + E_s I_{sy}$$
$$= 210 \times 276.7 \times 10^6 + 0.6 \times 27.2 \times 4856 \times 10^6 + 210 \times 75.36 \times 10^6$$
$$= 1.53 \times 10^{11}\,\text{kN mm}^2$$

$$(EI)_{eff,z} = E_a I_{az} + 0.6E_{c,eff}I_{cz} + E_s I_{sz}$$
$$= 210 \times 90.6 \times 10^6 + 0.6 \times 27.2 \times 5042 \times 10^6 + 210 \times 75.36 \times 10^6$$
$$= 1.17 \times 10^{11}\,\text{kN mm}^2$$

## Elastic critical normal force

$$N_{cry} = \frac{\pi^2 (EI)_{eff,y}}{L_y^2} = \frac{\pi^2 \times 1.53 \times 10^{11}}{4^2 \times 10^6} = 94,200 \, kN$$

$$N_{crz} = \frac{\pi^2 (EI)_{eff,z}}{L_z^2} = \frac{\pi^2 \times 1.17 \times 10^{11}}{4^2 \times 10^6} = 71,900 \, kN$$

The characteristic value of the plastic resistance to the axial load is:

$$
\begin{aligned}
N_{pl,Rk} &= A_a f_y + 0.85 A_c f_{ck} + A_s f_{sk} \\
&= (15,020 \times 355 + 0.85 \times 232,468 \times 72 + 2512 \times 500) \times 10^3 \\
&= 20,815 \, kN
\end{aligned}
$$

## Relative slenderness ratio

$$\overline{\lambda_y} = \sqrt{\frac{N_{pl,Rk}}{N_{cry}}} = \sqrt{\frac{20,815}{94,200}} = 0.47$$

$$\overline{\lambda_z} = \sqrt{\frac{N_{pl,Rk}}{N_{crz}}} = \sqrt{\frac{20,815}{71,900}} = 0.54$$

## Buckling reduction factor

Buckling curve $b$ is applicable to axis $y$-$y$, and buckling curve $c$ is applicable to axis $z$-$z$ in accordance with EN 1994-1-1. The imperfection factor is taken as 0.34 for curve $b$ and 0.49 for curve $c$. According to EN 1993-1-1, the factor is:

$$
\begin{aligned}
\Phi_y &= 0.5 \left( 1 + \alpha \left( \overline{\lambda_y} - 0.2 \right) + \overline{\lambda_y^2} \right) \\
&= 0.5 \times [1 + 0.34 \times (0.47 - 0.2) + 0.47^2] = 0.66
\end{aligned}
$$

$$
\begin{aligned}
\Phi_z &= 0.5 \left( 1 + \alpha \left( \overline{\lambda_z} - 0.2 \right) + \overline{\lambda_z^2} \right) \\
&= 0.5 \times [1 + 0.49 \times (0.54 - 0.2) + 0.54^2] = 0.73
\end{aligned}
$$

The reduction factor for column buckling is:

$$\chi_y = \frac{1}{\Phi_y + \sqrt{\Phi_y^2 - \overline{\lambda}_y^2}} = \frac{1}{0.66 + \sqrt{0.66^2 - 0.47^2}} = 0.90$$

$$\chi_z = \frac{1}{\Phi_z + \sqrt{\Phi_z^2 - \overline{\lambda}_z^2}} = \frac{1}{0.73 + \sqrt{0.73^2 - 0.54^2}} = 0.82$$

## Buckling resistance

The minor axis is the more critical, so

$$N_{b,Rd} = \min(\chi_y; \chi_z) N_{pl,Rd}$$
$$= 0.82 \times 15,909 = 13,045 \, \text{kN} > N_{Ed} = 9000 \, \text{kN}$$

The buckling resistance of the SRC column is adequate.

## Interaction curve

### Point A $(0, N_{pl,Rd})$

The full cross-section is under compression without the bending moment.

$$M_A = 0$$
$$N_A = N_{pl,Rd} = 15,909 \text{ kN}$$

### Point B $(M_{pl,Rd}, 0)$

Assuming the neutral axis lies in the flange of the steel section $(h/2 - t_f < h_n \le h/2)$, the 2 reinforcement bar lies within the region $2h_n$, $A_{sn} = 628 \text{ mm}^2$, so,

Neutral axis in the flange, $h/2 - t_f < h_n < h/2$

$$
\begin{aligned}
h_n &= \frac{A_c \alpha_c f_{cd} - A_{sn}(2f_{sd} - \alpha_c f_{cd}) + (b - t_w)(h - 2t_f)(2f_{yd} - \alpha_c f_{cd})}{2b_c \alpha_c f_{cd} + 2b(2f_{yd} - \alpha_c f_{cd})} \\[6pt]
&= \frac{\begin{array}{c}232,468 \times 0.85 \times 48 - 628 \times (2 \times 435 - 0.85 \times 48) \\ +(307.4 - 12)(314.5 - 2 \times 18.7)(2 \times 355 - 0.85 \times 48)\end{array}}{2 \times 500 \times 0.85 \times 48 + 2 \times 307.4 \times (2 \times 355 - 0.85 \times 48)} \\[6pt]
&= 141 \, \text{mm}
\end{aligned}
$$

Hence,

$$h_n = 141 \text{mm} < \frac{h}{2} = 157 \text{mm}$$

The assumption for the plastic neutral axis is verified. The neutral axis lies in the flange of the steel section.

The plastic section moduli for the steel section, reinforcement, and concrete are:

$$W_{pa} = 1.958 \times 10^6 \text{ mm}^3$$

$$W_{ps} = \sum_{1}^{6} A_{si}[e_i] = 1884 \times 200 = 0.3768 \times 10^6 \text{ mm}^3$$

$$W_{pc} = \frac{b_c h_c^2}{4} - W_{pa} - W_{ps} = \frac{500 \times 500^2}{4} - 1.958 \times 10^6 - 0.3768 \times 10^6$$
$$= 28.915 \times 10^6 \text{ mm}^3$$

The plastic section moduli for the region of depth $2h_n$ are:

$$W_{psn} = 0 \text{ mm}^3$$

$$W_{pa,n} = bh_n^2 - \frac{(b-t_w)\left(h-2t_f^2\right)}{4}$$
$$= 307.4 \times 141^2 - \frac{(307.4-12)(314.5-2 \times 18.7)^2}{4} = 0.441 \times 10^6$$

$$W_{pcn} = b_c h_n^2 - W_{pan} - W_{psn} = 500 \times 141^2 - 0.441 \times 10^6 - 0$$
$$= 9.5 \times 10^6 \text{ mm}^3$$

The bending resistance at point B is determined from:

$$M_{pl,Rd} = (W_{pa} - W_{pa,n})f_{yd} + (W_{ps} - W_{ps,n})f_{sd} + 0.5\,(W_{pc} - W_{pc,n})\,\alpha_c f_{cd}$$
$$= (1.958 - 0.441) \times 355 + (0.3768 - 0) \times 435$$
$$+ 0.5 \times (28.915 - 9.5) \times 0.85 \times 48 = 1099 \text{ kNm}$$

## Point C ($M_{\text{pl,Rd}}$, $N_{\text{pm,Rd}}$)

The axial force is equal to the full cross-section compression resistance of concrete. The value is determined from:

$$N_{\text{pm,Rd}} = 0.85 A_c f_{cd} = 0.85 \times 232,468 \times 48 \times 10^{-3} = 9485\,\text{kN}$$

## Point D ($M_{\text{max,Rd}}$, $0.5 N_{\text{pm,Rd}}$)

The maximum moment resistance is determined from:

$$
\begin{aligned}
M_{\text{max,Rd}} &= f_{yd} W_{pa} + 0.5 \times 0.85 f_{cd} W_{pc} + f_{sd} W_{ps} \\
&= 355 \times 1.958 + 0.5 \times 0.85 \times 48 \times 28.915 + 435 \times 0.3768 \\
&= 1499\,\text{kNm}
\end{aligned}
$$

The relative information for plotting the interaction curve is shown in Table 2. Then, the interaction curve is plotted as shown in Figure 4.

### Check the resistance of steel-reinforced concrete column in combined compression and uniaxial bending

The effective flexural stiffness considering second-order effects is determined from:

$$
\begin{aligned}
(EI)_{\text{eff,II,y}} &= K_o (E_a I_{ay} + K_{e,II} E_{c,\text{eff}} I_{cy} + E_s I_{sy}) \\
&= 0.9 \times (210 \times 276.7 \times 10^6 + 0.5 \times 27.2 \\
&\quad \times 4856 \times 10^6 + 210 \times 75.36 \times 10^6) \\
&= 1.26 \times 10^{11}\,\text{kN mm}^2
\end{aligned}
$$

*Table 2* The resistance for interaction curve

| Point | Resistance to bending (kNm) | Resistance to compression (kN) |
|-------|------------------------------|--------------------------------|
| A | 0 | 15,909 |
| B | 1099 | 0 |
| C | 1099 | 9485 |
| D | 1499 | 4742 |

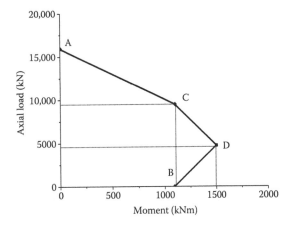

*Figure 4* Interaction curve for major axis.

Hence, the elastic critical force is:

$$N_{cr,y,eff} = \frac{\pi^2 (EI)_{eff,II,y}}{L_y^2} = \frac{\pi^2 \times 1.26 \times 10^{11}}{4^2 \times 10^6} = 77,400 \, \text{kN}$$

The result is less than $10N_{Ed}$ for the major axis, so the second-order effects must be considered for the moment from the first-order analysis and the moment from imperfection.

The member imperfection for the major axis according to EN1994-1-1 is:

$$e_{0,y} = L/200 = 20 \, \text{mm}$$

For the major axis, the midlength bending moments due to $N_{Ed}$ and imperfection are calculated by:

$$N_{Ed} e_{0,y} = 9000 \times 0.02 = 180 \, \text{kNm}$$

According to EN 1994-1-1, the factor $\beta$ is equal to 1.0 for the bending moment from the member imperfection. Then, the amplification factor is:

$$k_{imp,y} = \frac{\beta}{1 - N_{Ed}/N_{cr,y,eff}} = \frac{1.0}{1 - 9000/77,400} = 1.13$$

For the first-order bending moment, $M_{y,top} = 200$ kNm, $M_{y,bot} = 300$ kNm, so the ratio of the end moment is:

$$r = 200/300 = 0.667$$

The factor $\beta$ is:

$$\beta = \max{(0.66 + 0.44\ r;\ 0.44)} = 0.95$$

Thus, the amplification factor is:

$$k_y = \frac{\beta}{1 - N_{Ed}/N_{cr,y,eff}} = \frac{0.95}{1 - 9000/77,400} = 1.08$$

Hence, the design moment considering second-order effects is:

$$M_{y,Ed} = k_y M_{y,Ed,top} + k_{imp,y} N_{Ed} e_{0,z} = 1.08 \times 300 + 1.13 \times 200$$
$$= 550\,\text{kNm}$$

For $0.5 N_{pm,Rd} < N_{Ed} \leq N_{pm,Rd}$, the factor is determined from:

$$\mu_d = 1 + \frac{2(N_{pm,Rd} - N_{Ed})}{N_{pm,Rd}} \left( \frac{M_{max,Rd}}{M_{pl,Rd}} - 1 \right)$$
$$= 1 + \frac{2 \times (9485 - 9000)}{9485} \left( \frac{1499}{1099} - 1 \right) = 1.04$$

Thus,

$$\frac{M_{y,Ed}}{M_{pl,N,y,Rd}} = \frac{M_{y,Ed}}{\mu_d M_{pl,y,Rd}} = \frac{550}{1.04 \times 1099} = 0.48 < 0.9$$

The resistance of the SRC column to compression and uniaxial bending is satisfied.

## Steel-reinforced concrete column with high-strength steel

The steel grade S550 is used in this design example. Other design data are same as in Section "Steel-reinforced concrete column with normal-strength material," such as loading; column length; concrete strength; and dimensions of the SRC column cross-section, steel section, and reinforcement.

### Design strengths and modulus

According to Tables 2.12 and 2.13 in the companion book, for the strain compatibility between steel and concrete, the two materials (concrete class C50/60 and steel grade S550) are not strain compatible. Therefore, high-strength concrete reaches its peak strain much earlier than the yield strain of steel. This implies that concrete will fail earlier than steel, resulting in a partial utilization of steel strength. Using the strain-compatibility method, the strength of steel is limited to the stress corresponding to the crushing strain of concrete. The confinement effect from the lateral hoops and steel section is considered as follows.

Longitudinal reinforcement ratio:

$$\rho_s = \frac{A_s}{A_c} = 1.1\%$$

The confinement effective coefficient:

$$k_e = \frac{\left(1 - \sum_{i=1}^{n} ((b_i)^2 / 6 b_c h_c)\right)(1 - (s/2b_c))(1 - (s/2h_c))}{1 - \rho_s} = 0.5135$$

The effective volume ratio of the hoops is:

$$\rho_{se} = k_e \rho_{s,h} = 0.5135 \times 0.2\% = 0.1\%$$

The real stress of the hoops is:

$$\kappa = \frac{f_{c,u}}{\rho_{se} E_s \varepsilon_c} = \frac{50}{0.001 \times 210 \times 0.0025} = 100$$

$$f_{r,h} = \max\left(\frac{0.25 f_{c,u}}{\rho_{se}(\kappa - 10)}; 0.43 \varepsilon_c E_s\right)$$

$$= \max\left(\frac{0.25 \times 50}{0.001(100 - 10)}; 0.43 \times 0.0025 \times 210,000\right) = 223\,\text{N/mm}^2$$

The effective lateral confining pressure for PCC from the hoops is:

$$f_{l,p} = \rho_{se}f_{r,h} = 0.001 \times 223 = 0.223\,\text{N/mm}^2$$

The strain of PCC is:

$$\varepsilon_{c,p} = \left[1+35\left(\frac{f_{l,p}}{f_{c,u}}\right)^{1.2}\right]\varepsilon_c = \left[1+35\left(\frac{0.223}{50}\right)^{1.2}\right] \times 0.0025 = 0.0026$$

Factor:

$$k'_e = \frac{A_{c,f} - A_{c,r}}{A_{c,f}} = \frac{40,927 - 12,797}{40,927} = 0.69$$

Factor:

$$k_a = \frac{t_f^2}{3l^2} = 0.0053$$

The effective lateral confining pressure from the steel section is:

$$f_{l,s} = k'_e k_a f_{r,y} = 1.9\,\text{N/mm}^2$$

The effective lateral confining stress for HCC is:

$$f_{l,h} = f_{l,p} + f_{l,s} = 0.223 + 1.9 = 2.123\,\text{N/mm}^2$$

The strain of HCC is:

$$\varepsilon_{c,p} = \left[1+35\left(\frac{f_{l,h}}{f_{c,u}}\right)^{1.2}\right]\varepsilon_c = \left[1+35\left(\frac{2.123}{50}\right)^{1.2}\right] \times 0.0025 = 0.004$$

To ensure the yield strain of steel is less than the compressive strain of concrete, the maximum steel strength can be determined accordingly.

The real stress of the steel flange in partially confined concrete is:

$$f_{r,f} = \varepsilon_{c,p}E_a = 0.0026 \times 210,000 = 546\,\text{N/mm}^2$$

The real stress of the steel web in highly confined concrete is:

$$f_{r,w} = \min(\varepsilon_{c,h}E_a; f_y) = \min(0.004 \times 210,000; 550) = 550 \,\text{N/mm}^2$$

The steel strength in partially confined concrete is lower than the yield strength of steel, 550 N/mm$^2$. The confinement pressure is insufficient to ensure the utilization of steel's full strength. A higher confinement level is needed. Thus, the conservative value of the steel flange is taken as the steel strength in the following design.

Then, the design strengths of steel is:

$$f_{yd} = \frac{f_y}{\gamma_M} = \frac{546}{1.0} = 546 \,\text{N/mm}^2$$

### Cross-sectional areas and second moments of area

$A_a = 15,020 \text{ mm}^2$, $A_s = 2512 \text{ mm}^2$, $A_c = 232,468 \text{ mm}^2$
$I_{ay} = 276.7 \times 10^6 \text{ mm}^4$, $I_{az} = 90.6 \times 10^6 \text{ mm}^4$
$I_{sy} = 75.36 \times 10^6 \text{ mm}^4$, $I_{sz} = 75.36 \times 10^6 \text{ mm}^4$
$I_{cy} = 4856 \times 10^6 \text{ mm}^4$, $I_{cz} = 5042 \times 10^6 \text{ mm}^4$

### Check the steel contribution factor

The design plastic resistance of the composite cross-section in compression is:

$$\begin{aligned}
N_{pl,Rd} &= A_a f_{yd} + 0.85 A_c f_{cd} + A_s f_{sd} \\
&= (15,020 \times 546 + 0.85 \times 232,468 \times 33.3 + 2512 \times 435) \times 10^{-3} \\
&= 15,880 \,\text{kN}
\end{aligned}$$

$$\delta = \frac{A_a f_{yd}}{N_{pl,Rd}} = \frac{15,020 \times 546 \times 10^{-3}}{15,880} = 0.52 < 0.9$$

### Long-term effects

The creep coefficient is:

$$\varphi_t = 1.53$$

### Elastic modulus of concrete considering long-term effects

The modulus of elasticity of concrete $E_{cm}$ is reduced to the value $E_{c,eff}$:

$$E_{c,eff} = \frac{E_{cm}}{1 + (N_{G,Ed}/N_{Ed})\varphi_t} = 22.2\,\text{kN/mm}^2$$

### Effective flexural stiffness of cross-section

The effective elastic flexural stiffness taking account of the long-term effects is:

$$(EI)_{eff,y} = E_a I_{ay} + 0.6 E_{c,eff} I_{cy} + E_s I_{sy} = 1.38 \times 10^{11}\,\text{kN mm}^2$$

$$(EI)_{eff,z} = E_a I_{az} + 0.6 E_{c,eff} I_{cz} + E_s I_{sz} = 1.01 \times 10^{11}\,\text{kN mm}^2$$

### Elastic critical normal force

$$N_{cry} = \frac{\pi^2 (EI)_{eff,y}}{L_y^2} = 85,100\,\text{kN}$$

$$N_{crz} = \frac{\pi^2 (EI)_{eff,z}}{L_z^2} = 62,500\,\text{kN}$$

The characteristic value of the plastic resistance to the axial load is:

$$
\begin{aligned}
N_{pl,Rk} &= A_a f_y + 0.85 A_c f_{ck} + A_s f_{sk} \\
&= (15,020 \times 546 + 0.85 \times 232,468 \times 50 + 2512 \times 500) \times 10^3 \\
&= 19,337\,\text{kN}
\end{aligned}
$$

### Relative slenderness ratio

$$\overline{\lambda}_y = \sqrt{\frac{N_{pl,Rk}}{N_{cry}}} = \sqrt{\frac{19,337}{85,100}} = 0.48$$

$$\overline{\lambda}_z = \sqrt{\frac{N_{pl,Rk}}{N_{crz}}} = \sqrt{\frac{19,337}{62,500}} = 0.56$$

## Buckling reduction factor

Buckling curve $b$ is applicable to axis $y$-$y$, and buckling curve $c$ is applicable to axis $z$-$z$ in accordance with EN 1994-1-1. The imperfection factor is taken as 0.34 for curve $b$ and 0.49 for curve $c$. According to EN 1993-1-1, the factor is:

$$\Phi_y = 0.5\left(1 + \alpha\left(\overline{\lambda}_y - 0.2\right) + \overline{\lambda}_y^2\right)$$
$$= 0.5 \times [1 + 0.34 \times (0.48 - 0.2) + 0.48^2] = 0.66$$

$$\Phi_z = 0.5\left(1 + \alpha\left(\overline{\lambda}_z - 0.2\right) + \overline{\lambda}_z^2\right)$$
$$= 0.5 \times [1 + 0.49 \times (0.56 - 0.2) + 0.56^2] = 0.74$$

The reduction factor for column buckling is:

$$\chi_y = \frac{1}{\Phi_y + \sqrt{\Phi_y^2 - \overline{\lambda}_y^2}} = \frac{1}{0.66 + \sqrt{0.66^2 - 0.48^2}} = 0.89$$

$$\chi_z = \frac{1}{\Phi_z + \sqrt{\Phi_z^2 - \overline{\lambda}_z^2}} = \frac{1}{0.74 + \sqrt{0.74^2 - 0.56^2}} = 0.81$$

## Buckling resistance

The minor axis is the more critical, so

$$N_{b,Rd} = \min(\chi_y; \chi_z)N_{pl,Rd}$$
$$= 0.81 \times 15,880 = 12,863\,\text{kN} > N_{Ed} = 9000\,\text{kN}$$

The buckling resistance of the SRC column is adequate.

## Interaction curve

### Point A $(0, N_{pl,Rd})$

The full cross-section is under compression without the bending moment.

$$M_A = 0$$
$$N_A = N_{pl,Rd} = 15,880 \text{ kN}$$

### Point B $(M_{pl,Rd}, 0)$

Assuming the neutral axis lies in the web of the steel section ($h_n \leq h/2 - t_f$), the 2 reinforcement bar lies within the region $2h_n$, $A_{sn} = 628 \text{ mm}^2$, so,

$$
\begin{aligned}
h_n &= \frac{A_c 0.85 f_{cd} - A_{sn}(2f_{sd} - 0.85 f_{cd})}{2b_c 0.85 f_{cd} + 2t_w(2f_{yd} - 0.85 f_{cd})} \\
&= \frac{232,468 \times 0.85 \times 33.3 - 628 \times (2 \times 435 - 0.85 \times 33.3)}{2 \times 500 \times 0.85 \times 33.3 + 2 \times 12 \times (2 \times 546 - 0.85 \times 33.3)} = 113 \text{mm}
\end{aligned}
$$

Hence,

$$h_n = 113 \text{mm} < \frac{h}{2} - t_f = 138.6 \text{ mm}$$

The assumption for the plastic neutral axis is verified. The neutral axis lies in the web of the steel section.

The plastic section moduli for the steel section, reinforcement, and concrete are:

$$W_{pa} = 1.958 \times 10^6 \text{ mm}^3$$

$$W_{ps} = \sum_{1}^{6} A_{si}[e_i] = 1884 \times 200 = 0.3768 \times 10^6 \text{ mm}^3$$

$$
\begin{aligned}
W_{pc} &= \frac{b_c h_c^2}{4} - W_{pa} - W_{ps} = \frac{500 \times 500^2}{4} - 1.958 \times 10^6 - 0.3768 \times 10^6 \\
&= 28.915 \times 10^6 \text{ mm}^3
\end{aligned}
$$

The plastic section moduli for the region of depth $2h_n$ are:

$$W_{psn} = 0\,\text{mm}^3$$

$$W_{pan} = t_w h_n^2 = 12 \times 113^2 = 0.153 \times 10^6\,\text{mm}^3$$

$$W_{pcn} = b_c h_n^2 - W_{pan} - W_{psn} = 500 \times 113^2 - 0.153 \times 10^6 - 0$$
$$= 6.23 \times 10^6\,\text{mm}^3$$

The bending resistance at point B is determined from:

$$\begin{aligned}
M_{pl,Rd} &= (W_{pa} - W_{pa,n})f_{yd} + (W_{ps} - W_{ps,n})f_{sd} + 0.5(W_{pc} - W_{pc,n})\alpha_c f_{cd} \\
&= (1.958 - 0.153) \times 546 + (0.3768 - 0) \times 435 \\
&\quad + 0.5 \times (28.915 - 6.23) \times 0.85 \times 33.3 = 1470\,\text{kNm}
\end{aligned}$$

**Point C ($M_{pl,Rd}$, $N_{pm,Rd}$)**

The axial force is equal to the full cross-section compression resistance of concrete. The value is determined from:

$$N_{pm,Rd} = 0.85 A_c f_{cd} = 0.85 \times 232,468 \times 33.3 \times 10^{-3} = 6850\,\text{kN}$$

**Point D ($M_{max,Rd}$, $0.5 N_{pm,Rd}$)**

The maximum moment resistance is determined from:

$$\begin{aligned}
M_{max,Rd} &= f_{yd} W_{pa} + 0.5 \times 0.85 f_{cd} W_{pc} + f_{sd} W_{ps} \\
&= 546 \times 1.958 + 0.5 \times 0.85 \times 33.3 \times 28.915 + 435 \times 0.3768 \\
&= 1642\,\text{kNm}
\end{aligned}$$

The relative information for plotting the interaction curve is shown in Table 3. Then, the interaction curve is plotted as shown in Figure 5.

Table 3 The resistance for interaction curve

| Point | Resistance to bending (kNm) | Resistance to compression (kN) |
|-------|-----------------------------|--------------------------------|
| A | 0 | 15,880 |
| B | 1470 | 0 |
| C | 1470 | 6850 |
| D | 1642 | 3425 |

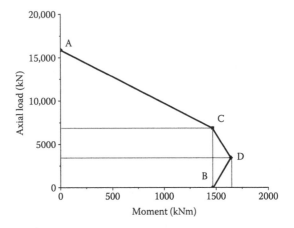

*Figure 5* Interaction curve for major axis.

## Check the resistance of steel-reinforced concrete column in combined compression and uniaxial bending

The effective flexural stiffness considering second-order effects is determined from:

$$(EI)_{\mathrm{eff,II,y}} = K_o(E_a I_{ay} + K_{e,II} E_{c,eff} I_{cy} + E_s I_{sy}) = 1.15 \times 10^{11}\,\mathrm{kN\,mm^2}$$

Hence, the elastic critical force is:

$$N_{\mathrm{cr,y,eff}} = \frac{\pi^2 (EI)_{\mathrm{eff,II,y}}}{L_y^2} = \frac{\pi^2 \times 1.15 \times 10^{11}}{4^2 \times 10^6} = 70,600\,\mathrm{kN}$$

The result is less than $10N_{\mathrm{Ed}}$ for the major axis, so the second-order effects must be considered for the moment from the first-order analysis and the moment from imperfection.

The member imperfection for the major axis according to EN1994-1-1 is:

$$e_{0,y} = L/200 = 20\,\mathrm{mm}$$

For the major axis, the midlength bending moments due to $N_{\mathrm{Ed}}$ and the imperfection are calculated by:

$$N_{Ed}e_{0,y} = 9000 \times 0.02 = 180 \, \text{kNm}$$

According to EN 1994-1-1, the factor $\beta$ is equal to 1.0 for the bending moment from the member imperfection. Then, the amplification factor is:

$$k_{\text{imp,y}} = \frac{\beta}{1 - N_{Ed}/N_{\text{cr,y,eff}}} = \frac{1.0}{1 - 9000/70,600} = 1.15$$

For the first-order bending moment, $M_{\text{y,top}} = 200 \, \text{kNm}, M_{\text{y,bot}} = 300 \, \text{kNm}$, so the ratio of the end moment is:

$$r = 200/300 = 0.667$$

Then, the factor $\beta$ is:

$$\beta = \max(0.66 + 0.44 \, r; \, 0.44) = 0.95$$

thus, the amplification factor is:

$$k_y = \frac{\beta}{1 - N_{Ed}/N_{\text{cr,y,eff}}} = \frac{0.95}{1 - 9000/70,600} = 1.09$$

Hence, the design moment considering the second-order effect is:

$$M_{\text{y,Ed}} = k_y M_{\text{y,Ed,top}} + k_{\text{imp,y}} N_{Ed} e_{0,z} = 1.09 \times 300 + 1.15 \times 200$$
$$= 557 \, \text{kNm}$$

For $N_{Ed} > N_{\text{pm,Rd}}$, the factor is determined from:

$$\mu_d = \frac{N_{\text{pl,Rd}} - N_{Ed}}{N_{\text{pl,Rd}} - N_{\text{pm,Rd}}} = \frac{15,880 - 9000}{15,880 - 6850} = 0.76$$

Thus,

$$\frac{M_{\text{y,Ed}}}{M_{\text{pl,N,y,Rd}}} = \frac{M_{\text{y,Ed}}}{\mu_d M_{\text{pl,y,Rd}}} = \frac{557}{0.76 \times 1470} = 0.5 < 0.9$$

So, the resistance of the SRC column to compression and uniaxial bending is satisfied.

## Steel-reinforced concrete column with high-strength materials

Steel grade **S550** and concrete class C90/105 are used in this design example. Other design data are same as in Section "Steel-reinforced concrete column with normal-strength material," such as loading; column length; dimensions of the SRC column cross-section, steel section, and reinforcement; and so on.

### Design strengths and modulus

According to Tables 2.12 and 2.13 in the companion book, for the strain compatibility between steel and concrete, the two materials (concrete class C90/105 and steel grade S550) are strain compatible, so the steel can reach its full strength when the composite concrete section reaches its ultimate strength without considering the confinement effect from the lateral hoops and steel section.

The effective compressive strength and elastic modulus of concrete C90/105 are:

$f_{ck} = 72$ N/mm$^2$; $f_{cd} = 48$ N/mm$^2$; $E_{cm} = 41.1$ GPa;

The design strength of steel is:

$f_y = 550$ N/mm$^2$; $f_{yd} = 550$ N/mm$^2$;

### Cross-sectional areas and second moments of area

$A_a = 15,020$ mm$^2$, $A_s = 2512$ mm$^2$, $A_c = 232,468$ mm$^2$
$I_{ay} = 276.7 \times 10^6$ mm$^4$, $I_{az} = 90.6 \times 10^6$ mm$^4$
$I_{sy} = 75.36 \times 10^6$ mm$^4$, $I_{sz} = 75.36 \times 10^6$ mm$^4$
$I_{cy} = 4856 \times 10^6$ mm$^4$, $I_{cz} = 5042 \times 10^6$ mm$^4$

### Check the steel contribution factor

The design plastic resistance of the composite cross-section in compression is:

$$N_{pl,Rd} = A_a f_{yd} + 0.85 A_c f_{cd} + A_s f_{sd}$$
$$= (15,020 \times 550 + 0.85 \times 232,468 \times 48 + 2512 \times 435) \times 10^{-3}$$
$$= 18,838\,kN$$

$$\delta = \frac{A_a f_{yd}}{N_{pl,Rd}} = \frac{15,020 \times 550 \times 10^{-3}}{18,838} = 0.44 < 0.9$$

## Long-term effects

The creep coefficient is:

$$\varphi_t = 1.14$$

## Elastic modulus of concrete considering long-term effects

The modulus of elasticity of concrete $E_{cm}$ is reduced to the value $E_{c,eff}$:

$$E_{c,eff} = \frac{E_{cm}}{1 + (N_{G,Ed}/N_{Ed})\varphi_t} = 27.2 \, kN/mm^2$$

## Effective flexural stiffness of cross-section

The effective elastic flexural stiffness taking account of the long-term effects is:

$$(EI)_{eff,y} = E_a I_{ay} + 0.6 E_{c,eff} I_{cy} + E_s I_{sy} = 1.53 \times 10^{11} \, kN \, mm^2$$

$$(EI)_{eff,z} = E_a I_{az} + 0.6 E_{c,eff} I_{cz} + E_s I_{sz} = 1.17 \times 10^{11} \, kN \, mm^2$$

## Elastic critical normal force

$$N_{cry} = \frac{\pi^2 (EI)_{eff,y}}{L_y^2} = 94,200 \, kN$$

$$N_{crz} = \frac{\pi^2 (EI)_{eff,z}}{L_z^2} = 71,900 \, kN$$

The characteristic value of the plastic resistance to the axial load is:

$$\begin{aligned}
N_{pl,Rk} &= A_a f_y + 0.85 A_c f_{ck} + A_s f_{sk} \\
&= (15,020 \times 550 + 0.85 \times 232,468 \times 72 + 2512 \times 500) \times 10^3 \\
&= 23,744 \, kN
\end{aligned}$$

### Relative slenderness ratio

$$\overline{\lambda}_y = \sqrt{\frac{N_{pl,Rk}}{N_{cry}}} = \sqrt{\frac{23,744}{94,200}} = 0.50$$

$$\overline{\lambda}_z = \sqrt{\frac{N_{pl,Rk}}{N_{crz}}} = \sqrt{\frac{23,744}{71,900}} = 0.57$$

### Buckling reduction factor

Buckling curve $b$ is applicable to axis $y$-$y$, and buckling curve $c$ is applicable to axis $z$-$z$ in accordance with EN 1994-1-1. The imperfection factor is taken as 0.34 for curve $b$ and 0.49 for curve $c$. According to EN 1993-1-1, the factor is:

$$\Phi_y = 0.5\left(1+\alpha\left(\overline{\lambda}_y-0.2\right)+\overline{\lambda}_y^2\right)$$
$$= 0.5\times[1+0.34\times(0.50-0.2)+0.50^2] = 0.68$$

$$\Phi_z = 0.5\left(1+\alpha\left(\overline{\lambda}_z-0.2\right)+\overline{\lambda}_z^2\right)$$
$$= 0.5\times[1+0.49\times(0.57-0.2)+0.57^2] = 0.76$$

The reduction factor for column buckling is:

$$\chi_y = \frac{1}{\Phi_y+\sqrt{\Phi_y^2-\overline{\lambda}_y^2}} = \frac{1}{0.68+\sqrt{0.68^2-0.50^2}} = 0.88$$

$$\chi_z = \frac{1}{\Phi_z+\sqrt{\Phi_z^2-\overline{\lambda}_z^2}} = \frac{1}{0.76+\sqrt{0.76^2-0.57^2}} = 0.80$$

### Buckling resistance

The minor axis is the more critical, so,

$$N_{b,Rd} = \min(\chi_y;\chi_z)N_{pl,Rd}$$
$$= 0.80\times18,838 = 15,070\,\text{kN} > N_{Ed} = 9000\,\text{kN}$$

The buckling resistance of the SRC column is adequate.

## Interaction curve

### Point A $(0, N_{pl,Rd})$

The full cross-section is under compression without the bending moment.

$$M_A = 0$$
$$N_A = N_{pl,Rd} = 18,838 \text{ kN}$$

### Point B $(M_{pl,Rd}, 0)$

Assuming the neutral axis lies in the web of the steel section ($h_n \leq h/2 - t_f$), the 2 reinforcement bar lies within the region $2h_n$, $A_{sn} = 628 \text{ mm}^2$, so,

$$
\begin{aligned}
h_n &= \frac{A_c 0.85 f_{cd} - A_{sn}(2f_{sd} - 0.85f_{cd})}{2b_c 0.85 f_{cd} + 2t_w(2f_{yd} - 0.85f_{cd})} \\
&= \frac{23,2468 \times 0.85 \times 48 - 628 \times (2 \times 435 - 0.85 \times 48)}{2 \times 500 \times 0.85 \times 48 + 2 \times 12 \times (2 \times 550 - 0.85 \times 48)} = 136 \text{ mm}
\end{aligned}
$$

Hence,

$$h_n = 136 \text{ mm} < \frac{h}{2} - t_f = 138.6 \text{ mm}$$

The assumption for the plastic neutral axis is verified. The neutral axis lies in the web of the steel section.

The plastic section moduli for the steel section, reinforcement, and concrete are:

$$W_{pa} = 1.958 \times 10^6 \text{ mm}^3$$

$$W_{ps} = \sum_{1}^{6} A_{si}[e_i] = 1884 \times 200 = 0.3768 \times 10^6 \text{ mm}^3$$

$$
\begin{aligned}
W_{pc} &= \frac{b_c h_c^2}{4} - W_{pa} - W_{ps} = \frac{500 \times 500^2}{4} - 1.958 \times 10^6 - 0.3768 \times 10^6 \\
&= 28.915 \times 10^6 \text{ mm}^3
\end{aligned}
$$

The plastic section moduli for the region of depth $2h_n$ are:

$$W_{psn} = 0\,\mathrm{mm}^3$$

$$W_{pan} = t_w h_n^2 = 12 \times 136^2 = 0.222 \times 10^6\,\mathrm{mm}^3$$

$$W_{pcn} = b_c h_n^2 - W_{pan} - W_{psn} = 500 \times 136^2 - 0.222 \times 10^6 - 0$$
$$= 9.026 \times 10^6\,\mathrm{mm}^3$$

The bending resistance at point B is determined from:

$$M_{pl,Rd} = (W_{pa} - W_{pa,n})f_{yd} + (W_{ps} - W_{ps,n})f_{sd} + 0.5(W_{pc} - W_{pc,n})\alpha_c f_{cd}$$
$$= (1.958 - 0.222) \times 550 + (0.3768 - 0) \times 435$$
$$+ 0.5 \times (28.915 - 9.026) \times 0.85 \times 48 = 1524\,\mathrm{kNm}$$

## Point C ($M_{pl,Rd}$, $N_{pm,Rd}$)

The axial force is equal to the full cross-section compression resistance of concrete. The value is determined from:

$$N_{pm,Rd} = 0.85 A_c f_{cd} = 0.85 \times 232,468 \times 48 \times 10^{-3} = 9485\,\mathrm{kN}$$

## Point D ($M_{max,Rd}$, $0.5 N_{pm,Rd}$)

The maximum moment resistance is determined from:

$$M_{max,Rd} = f_{yd} W_{pa} + 0.5 \times 0.85 f_{cd} W_{pc} + f_{sd} W_{ps}$$
$$= 550 \times 1.958 + 0.5 \times 0.85 \times 48 \times 28.915 + 435 \times 0.3768$$
$$= 1723\,\mathrm{kNm}$$

The relative information for plotting the interaction curve is shown in Table 4. Then, the interaction curve is plotted as shown in Figure 6.

Table 4 The resistance for interaction curve

| Point | Resistance to bending (kNm) | Resistance to compression (kN) |
|-------|-----------------------------|--------------------------------|
| A | 0 | 18,838 |
| B | 1524 | 0 |
| C | 1524 | 9485 |
| D | 1723 | 4742 |

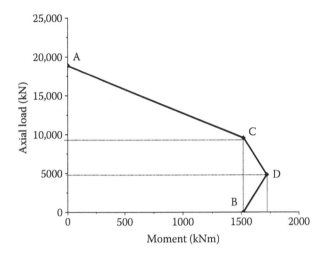

*Figure 6* Interaction curve for major axis.

## Check the resistance of steel-reinforced concrete column in combined compression and uniaxial bending

The effective flexural stiffness considering second-order effects is determined from:

$$(EI)_{\text{eff,II,y}} = K_o(E_aI_{\text{ay}} + K_{\text{e,II}}E_{\text{c,eff}}I_{\text{cy}} + E_sI_{\text{sy}}) = 1.26 \times 10^{11}\,\text{kN\,mm}^2$$

Hence, the elastic critical force is:

$$N_{\text{cr,y,eff}} = \frac{\pi^2(EI)_{\text{eff,II,y}}}{L_y^2} = \frac{\pi^2 \times 1.26 \times 10^{11}}{4^2 \times 10^6} = 77,400\,\text{kN}$$

The result is less than $10N_{\text{Ed}}$ for the major axis, so the second-order effects must be considered for the moment from the first-order analysis and the moment from the imperfection.

The member imperfection for the major axis according to EN1994-1-1 is:

$$e_{0,y} = L/200 = 20\,\text{mm}$$

For the major axis, the midlength bending moments due to $N_{Ed}$ and the imperfection are calculated by:

$$N_{Ed}e_{0,y} = 9000 \times 0.02 = 180 \, \text{kN m}$$

According to EN 1994-1-1, the factor $\beta$ is equal to 1.0 for the bending moment from the member imperfection. Then, the amplification factor is:

$$k_{\text{imp,y}} = \frac{\beta}{1 - N_{Ed}/N_{\text{cr,y,eff}}} = \frac{1.0}{1 - 9000/77,400} = 1.13$$

For the first-order bending moment, $M_{y,top} = 200$ kNm, $M_{y,bot} = 300$ kNm, so the ratio of the end moment is:

$$r = 200/300 = 0.667$$

Then, the factor $\beta$ is:

$$\beta = \max{(0.66 + 0.44 \, r; \, 0.44)} = 0.95$$

Thus, the amplification factor is:

$$k_y = \frac{\beta}{1 - N_{Ed}/N_{\text{cr,y,eff}}} = \frac{0.95}{1 - 9000/77,400} = 1.08$$

Hence, the design moment considering second-order effects is:

$$M_{y,Ed} = k_y M_{y,Ed,top} + k_{\text{imp,y}} N_{Ed} e_{0,z} = 1.08 \times 300 + 1.13 \times 200$$
$$= 550 \, \text{kNm}$$

For $0.5 N_{\text{pm,Rd}} < N_{Ed} \leq N_{\text{pm,Rd}}$, the factor is determined from:

$$
\begin{aligned}
\mu_d &= 1 + \frac{2\left(N_{\text{pm,Rd}} - N_{Ed}\right)}{N_{\text{pm,Rd}}}\left(\frac{M_{\text{max,Rd}}}{M_{\text{pl,Rd}}} - 1\right) \\
&= 1 + \frac{2 \times (9485 - 9000)}{9485}\left(\frac{1723}{1524} - 1\right) = 1.01
\end{aligned}
$$

Thus,

$$\frac{M_{y,Ed}}{M_{\text{pl,N,y,Rd}}} = \frac{M_{y,Ed}}{\mu_d M_{\text{pl,y,Rd}}} = \frac{550}{1.01 \times 1524} = 0.36 < 0.9$$

So, the resistance of the SRC column to compression and uniaxial bending is satisfied.

## STEEL-REINFORCED CONCRETE COLUMN WITH DIFFERENT DEGREE OF CONFINEMENT

A comparison between the original and alternative designs of the SRC column was conducted to study the effect of confinement on the strength of the column.

### Original design

Figure 7 shows the original design of SRC columns from a practical project. In the original design, the concrete cylinder strength was 50 MPa and the yield strengths of the steel and rebar were 355 and 500 MPa, respectively. The diameter and spacing of the transverse reinforcement (hoops) were 10 and 200 mm, respectively. The diameter of the longitudinal reinforcement was 25 mm.

### High-strength concrete

High-strength concrete (90 MPa) was proposed as an alternative design. Thus, the concrete and reinforcement bar amounts are reduced and the arrangement is revised. As a result, the dimension of the column is reduced, as shown in Figure 8.

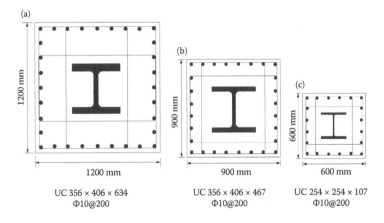

*Figure 7* Original design (S355 steel and C50/60 concrete). (a) N1, (b) N2, (c) N3.

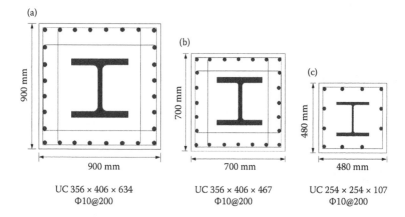

Figure 8 Alternative design with S355 steel and C90/105 concrete. (a) normal strength steel and High strength concrete (NH)1, (b) NH2, (c) NH3.

## High-strength steel and high-strength concrete

High-strength concrete (90 MPa) with high-strength steel (690 MPa) was proposed as an alternative design. Thus, the concrete and steel amounts were reduced and their arrangement was revised, as shown in Figure 9. To develop the full strength of steel, the spacing of the transverse reinforcement was made smaller than the original design

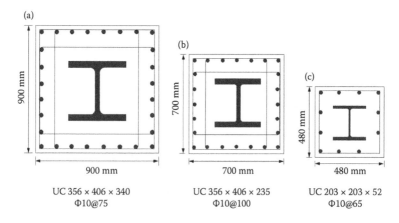

Figure 9 Alternative design with S690 steel and C90/105 concrete. (a) HH1, (b) HH2, (c) HH3.

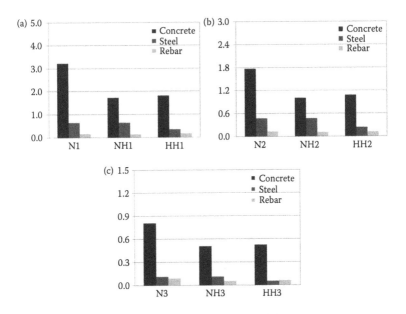

*Figure 10* Comparison of material amounts. (a) N1,NH1,HH1, (b) N2,NH2,HH2, (c) N3,NH3,HH3.

according to the equations from the modified confinement model. The smaller spacing of the hoops can lead to congestion, which is a practical concern that needs to be addressed.

Figure 10 shows the comparison of the concrete, steel, and rebar amounts of different cross-sections. There is a reduction of 47% (1.5 tons per meter) in the amount of concrete by adopting high-strength concrete C90/105. There is a reduction of 46% (0.3 ton per meter) in the amount of structural steel by adopting high-strength steel S690. However, the transverse reinforcement amount increases, as the spacing of the hoops is smaller than the original design. The transverse reinforcement increases by 200% (0.03 ton per meter) by adopting an S690 steel section, as shown in Figure 10a. Similarly, for original cross-sections N2 and N3, similar results can also be obtained. The concrete and steel amounts are reduced significantly by adopting high-strength material, but the amount of transverse reinforcement is increased by adopting an S690 steel section. Nevertheless, the total amount of concrete and steel (structural and reinforcing steel) is reduced significantly.

Figure 11 shows the effect of the spacing and diameter of links on the steel stress of the three types of SRC columns (high strength steel

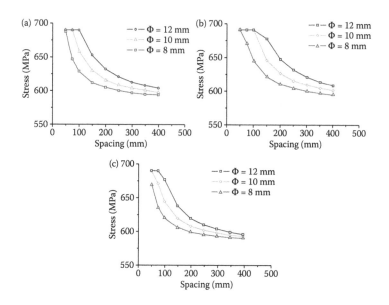

*Figure 11*  Effect of spacing and diameter of links. (a) HH1, (b) HH2, (c) HH3.

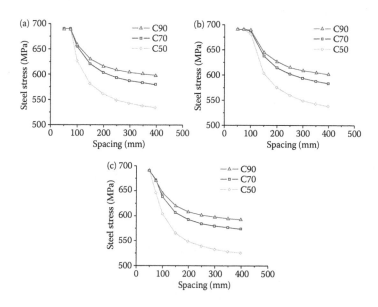

*Figure 12*  Effect of spacing of links and concrete strength. (a) HH1, (b) HH2, (c) HH3.

and high strength concrete [HH]1, HH2, and HH3). It indicates that the real stress of steel increases with the increase in the diameter of the transverse reinforcement when the spacing of links takes a certain value. The steel stress increases with the decreasing of the spacing of links due to insufficient lateral confinement, which has been verified in parametric study. For the three types of SRC columns, if the spacing of the hoops is same as the original design (200 mm); the real stress of steel grade S690 is about 600–650 MPa depending on the diameter of the hoops. For a large spacing of 400 mm, the real stress of steel grade S690 is 590–610 MPa depending on the diameter of the links and the dimensions of the column.

Concrete strength also has an effect on the stress in steel according to the strain-compatibility method because the strains of different concrete classes are different. Figure 12 shows the effect of the spacing of links and concrete cylinder strength on the real stress in steel. It is indicated that the real stress of steel increases with the increasing of the concrete cylinder strength for these three types of SRC columns. The stress in the steel section increases according to the stress–strain relationship of steel even though the lateral confinement pressure is the same. Therefore, for the same arrangement of the reinforcement bar, the stress in steel can reach a higher value when a higher-strength concrete is used in the SRC columns.

# Appendix A: Design resistance of shear connectors

Table A.1 Design resistance of shear connectors

| Dimension of connectors | | Design resistance of shear studs $P_{Rd}$ (kN) | | | | | | | | | | | |
|---|---|---|---|---|---|---|---|---|---|---|---|---|---|
| d (mm) | $h_{sc}$ (mm) | C20/25 | C25/30 | C30/37 | C35/45 | C40/50 | C45/55 | C50/60 | C55/67 | C60/75 | C70/85 | C80/95 | C90/105 |
| 16 | 50 | 37.3 | 43.1 | 48.8 | 53.5 | 57.9 | 57.9 | 57.9 | 57.9 | 57.9 | 57.9 | 57.9 | 57.9 |
| | 75 | 45.2 | 52.3 | 57.9 | 57.9 | 57.9 | 57.9 | 57.9 | 57.9 | 57.9 | 57.9 | 57.9 | 57.9 |
| | 100 | 45.2 | 52.3 | 57.9 | 57.9 | 57.9 | 57.9 | 57.9 | 57.9 | 57.9 | 57.9 | 57.9 | 57.9 |
| | 125 | 45.2 | 52.3 | 57.9 | 57.9 | 57.9 | 57.9 | 57.9 | 57.9 | 57.9 | 57.9 | 57.9 | 57.9 |
| | 150 | 45.2 | 52.3 | 57.9 | 57.9 | 57.9 | 57.9 | 57.9 | 57.9 | 57.9 | 57.9 | 57.9 | 57.9 |
| | 175 | 45.2 | 52.3 | 57.9 | 57.9 | 57.9 | 57.9 | 57.9 | 57.9 | 57.9 | 57.9 | 57.9 | 57.9 |
| | 200 | 45.2 | 52.3 | 57.9 | 57.9 | 57.9 | 57.9 | 57.9 | 57.9 | 57.9 | 57.9 | 57.9 | 57.9 |
| | 225 | 45.2 | 52.3 | 57.9 | 57.9 | 57.9 | 57.9 | 57.9 | 57.9 | 57.9 | 57.9 | 57.9 | 57.9 |
| | 250 | 45.2 | 52.3 | 57.9 | 57.9 | 57.9 | 57.9 | 57.9 | 57.9 | 57.9 | 57.9 | 57.9 | 57.9 |
| 19 | 50 | 46.3 | 53.6 | 60.5 | 66.4 | 72.0 | 77.4 | 81.7 | 81.7 | 81.7 | 81.7 | 81.7 | 81.7 |
| | 75 | 63.1 | 73.0 | 81.2 | 81.7 | 81.7 | 81.7 | 81.7 | 81.7 | 81.7 | 81.7 | 81.7 | 81.7 |
| | 100 | 63.8 | 73.7 | 81.7 | 81.7 | 81.7 | 81.7 | 81.7 | 81.7 | 81.7 | 81.7 | 81.7 | 81.7 |
| | 125 | 63.8 | 73.7 | 81.7 | 81.7 | 81.7 | 81.7 | 81.7 | 81.7 | 81.7 | 81.7 | 81.7 | 81.7 |
| | 150 | 63.8 | 73.7 | 81.7 | 81.7 | 81.7 | 81.7 | 81.7 | 81.7 | 81.7 | 81.7 | 81.7 | 81.7 |
| | 175 | 63.8 | 73.7 | 81.7 | 81.7 | 81.7 | 81.7 | 81.7 | 81.7 | 81.7 | 81.7 | 81.7 | 81.7 |
| | 200 | 63.8 | 73.7 | 81.7 | 81.7 | 81.7 | 81.7 | 81.7 | 81.7 | 81.7 | 81.7 | 81.7 | 81.7 |
| | 225 | 63.8 | 73.7 | 81.7 | 81.7 | 81.7 | 81.7 | 81.7 | 81.7 | 81.7 | 81.7 | 81.7 | 81.7 |
| | 250 | 63.8 | 73.7 | 81.7 | 81.7 | 81.7 | 81.7 | 81.7 | 81.7 | 81.7 | 81.7 | 81.7 | 81.7 |

(Continued)

Table A.1 (Continued) Design resistance of shear connectors

| Dimension of connectors | | Design resistance of shear studs $P_{Rd}$ (kN) | | | | | | | | | | | |
|---|---|---|---|---|---|---|---|---|---|---|---|---|---|
| d (mm) | $h_{sc}$ (mm) | C20/25 | C25/30 | C30/37 | C35/45 | C40/50 | C45/55 | C50/60 | C55/67 | C60/75 | C70/85 | C80/95 | C90/105 |
| | 275 | 63.8 | 73.7 | 81.7 | 81.7 | 81.7 | 81.7 | 81.7 | 81.7 | 81.7 | 81.7 | 81.7 | 81.7 |
| | 300 | 63.8 | 73.7 | 81.7 | 81.7 | 81.7 | 81.7 | 81.7 | 81.7 | 81.7 | 81.7 | 81.7 | 81.7 |
| | 325 | 63.8 | 73.7 | 81.7 | 81.7 | 81.7 | 81.7 | 81.7 | 81.7 | 81.7 | 81.7 | 81.7 | 81.7 |
| | 350 | 63.8 | 73.7 | 81.7 | 81.7 | 81.7 | 81.7 | 81.7 | 81.7 | 81.7 | 81.7 | 81.7 | 81.7 |
| 22 | 50 | 56.0 | 64.7 | 73.1 | 80.2 | 87.0 | 93.5 | 100.0 | 106.3 | 109.5 | 109.5 | 109.5 | 109.5 |
| | 75 | 75.4 | 87.2 | 98.5 | 108.0 | 109.5 | 109.5 | 109.5 | 109.5 | 109.5 | 109.5 | 109.5 | 109.5 |
| | 100 | 85.5 | 98.9 | 109.5 | 109.5 | 109.5 | 109.5 | 109.5 | 109.5 | 109.5 | 109.5 | 109.5 | 109.5 |
| | 125 | 85.5 | 98.9 | 109.5 | 109.5 | 109.5 | 109.5 | 109.5 | 109.5 | 109.5 | 109.5 | 109.5 | 109.5 |
| | 150 | 85.5 | 98.9 | 109.5 | 109.5 | 109.5 | 109.5 | 109.5 | 109.5 | 109.5 | 109.5 | 109.5 | 109.5 |
| | 175 | 85.5 | 98.9 | 109.5 | 109.5 | 109.5 | 109.5 | 109.5 | 109.5 | 109.5 | 109.5 | 109.5 | 109.5 |
| | 200 | 85.5 | 98.9 | 109.5 | 109.5 | 109.5 | 109.5 | 109.5 | 109.5 | 109.5 | 109.5 | 109.5 | 109.5 |
| | 225 | 85.5 | 98.9 | 109.5 | 109.5 | 109.5 | 109.5 | 109.5 | 109.5 | 109.5 | 109.5 | 109.5 | 109.5 |
| | 250 | 85.5 | 98.9 | 109.5 | 109.5 | 109.5 | 109.5 | 109.5 | 109.5 | 109.5 | 109.5 | 109.5 | 109.5 |
| | 275 | 85.5 | 98.9 | 109.5 | 109.5 | 109.5 | 109.5 | 109.5 | 109.5 | 109.5 | 109.5 | 109.5 | 109.5 |
| | 300 | 85.5 | 98.9 | 109.5 | 109.5 | 109.5 | 109.5 | 109.5 | 109.5 | 109.5 | 109.5 | 109.5 | 109.5 |
| | 325 | 85.5 | 98.9 | 109.5 | 109.5 | 109.5 | 109.5 | 109.5 | 109.5 | 109.5 | 109.5 | 109.5 | 109.5 |
| | 350 | 85.5 | 98.9 | 109.5 | 109.5 | 109.5 | 109.5 | 109.5 | 109.5 | 109.5 | 109.5 | 109.5 | 109.5 |

(Continued)

Table A.1 (Continued) Design resistance of shear connectors

| Dimension of connectors | | Design resistance of shear studs $P_{Rd}$ (kN) | | | | | | | | | | | | | | |
|---|---|---|---|---|---|---|---|---|---|---|---|---|---|---|---|---|
| d (mm) | $h_{sc}$ (mm) | C20/25 | C25/30 | C30/37 | C35/45 | C40/50 | C45/55 | C50/60 | C55/67 | C60/75 | C70/85 | C80/95 | C90/105 |
| 25 | 50 | 66.3 | 76.6 | 86.6 | 94.9 | 102.9 | 110.7 | 118.3 | 125.8 | 133.1 | | | |
| | 75 | 88.3 | 102.1 | 115.4 | 126.5 | 137.3 | 141.4 | 141.4 | 141.4 | 141.4 | 141.4 | 141.4 | 141.4 |
| | 100 | 110.4 | 127.6 | 141.4 | 141.4 | 141.4 | 141.4 | 141.4 | 141.4 | 141.4 | 141.4 | 141.4 | 141.4 |
| | 125 | 110.4 | 127.6 | 141.4 | 141.4 | 141.4 | 141.4 | 141.4 | 141.4 | 141.4 | 141.4 | 141.4 | 141.4 |
| | 150 | 110.4 | 127.6 | 141.4 | 141.4 | 141.4 | 141.4 | 141.4 | 141.4 | 141.4 | 141.4 | 141.4 | 141.4 |
| | 175 | 110.4 | 127.6 | 141.4 | 141.4 | 141.4 | 141.4 | 141.4 | 141.4 | 141.4 | 141.4 | 141.4 | 141.4 |
| | 200 | 110.4 | 127.6 | 141.4 | 141.4 | 141.4 | 141.4 | 141.4 | 141.4 | 141.4 | 141.4 | 141.4 | 141.4 |
| | 225 | 110.4 | 127.6 | 141.4 | 141.4 | 141.4 | 141.4 | 141.4 | 141.4 | 141.4 | 141.4 | 141.4 | 141.4 |
| | 250 | 110.4 | 127.6 | 141.4 | 141.4 | 141.4 | 141.4 | 141.4 | 141.4 | 141.4 | 141.4 | 141.4 | 141.4 |
| | 275 | 110.4 | 127.6 | 141.4 | 141.4 | 141.4 | 141.4 | 141.4 | 141.4 | 141.4 | 141.4 | 141.4 | 141.4 |
| | 300 | 110.4 | 127.6 | 141.4 | 141.4 | 141.4 | 141.4 | 141.4 | 141.4 | 141.4 | 141.4 | 141.4 | 141.4 |
| | 325 | 110.4 | 127.6 | 141.4 | 141.4 | 141.4 | 141.4 | 141.4 | 141.4 | 141.4 | 141.4 | 141.4 | 141.4 |
| | 350 | 110.4 | 127.6 | 141.4 | 141.4 | 141.4 | 141.4 | 141.4 | 141.4 | 141.4 | 141.4 | 141.4 | 141.4 |

Note: $f_u = 450$ N/mm².

# Appendix B: Design chart

The design process for SRC column subjected to axial compression is summarized in the chart given in Figure B.1 whereas the design process for SRC column subjected to axial compression and bending is summarized in the chart given in Figure B.2.

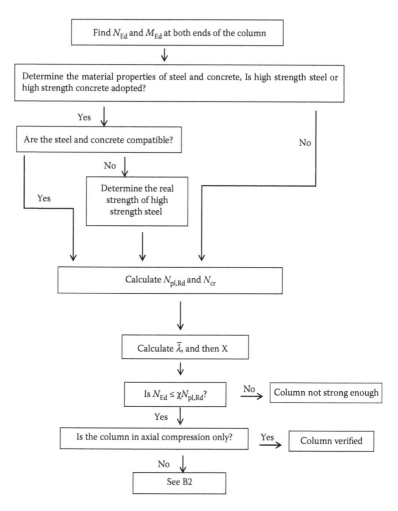

*Figure B.1* Design for SRC column subjected to axial compression.

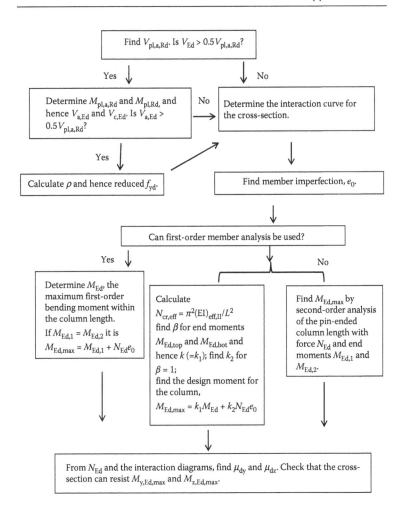

*Figure B.2* **Design for SRC column subjected to combined compression and bending.**

# Index